Geospatial Abduction

Paulo Shakarian • V.S. Subrahmanian

Geospatial Abduction

Principles and Practice

Foreword by Charles P. Otstott

 Springer

Paulo Shakarian
Kings Valley Drive
Bowie Maryland
USA
paulo@shakarian.net

V.S. Subrahmanian
University of Maryland - College Park
Potòmac Maryland
USA
vs@umiacs.umd.edu

ISBN 978-1-4899-9785-2   ISBN 978-1-4614-1794-1 (eBook)
DOI 10.1007/978-1-4614-1794-1
Springer New York Dordrecht Heidelberg London

Printed on acid-free paper

Springer is part of Springer Science+Business Media (www.springer.com)

To our wives, Jana and Mary

Foreword

Increasingly, US forces, multinational coalition troops, and UN peacekeeping forces worldwide are coming under asymmetric attacks including *Improvised Explosive Device* (IED) attacks. These attacks are carried out by adversaries who are outmatched in conventional military capabilities by opposing forces. However, such adversaries take advantage of their deep knowledge of the local terrain, ethno-social makeup of the areas in which they operate, and knowledge of local history and culture, to launch successful IED attacks.

This book is the first in the field of *geospatial abduction*—a mathematical and computational technique invented by the authors to solve a variety of problems including locating weapons caches that support IED attacks, identifying habitats that support hosts of disease-carrying viruses, identifying illegal drug labs and/or drug distribution centers, and solving problems related to locations of burglars and/or serial killers. Anyone with an interest in performing geospatial inferences and analytics involving these and other applications where one needs to identify locations based on a few observed clues, would find this book invaluable.

College Park, MD, August 2011

Charles P. Otstott
Lieutenant General US Army (Ret.)

Preface

There are numerous applications in which we make observations about events in time and space, and where we wish to infer one or more other locations that are causally related to the observations we made. Such applications span a number of fields and include finding habitats of hosts of disease carrying viruses (epidemiology), identifying locations where elusive animals such as tigers and leopards might live (wildlife conservation), identifying locations of illegal drug labs and distribution centers (criminology), and identifying the locations of weapons caches that facilitate improvised explosive device (IED) attacks on soldiers or civilians.

This book describes the technique of *geospatial abduction* in which we can use knowledge about the locations of observations, as well as certain application specific properties, to make the types of inferences needed to enable applications such as those listed above.

The book is intended as a research monograph on this emerging new field. The book provides a detailed technical definition of many forms of geospatial abduction, as well as many different theoretical results, as well as exact algorithms and approximation algorithms and heuristics to solve geospatial abduction problems. In order to facilitate readability, a number of examples are used throughout the book to illustrate basic concepts and algorithms. Moreover, we discuss in detail, a specific application to finding weapons caches that terrorist groups use to launch IED attacks.

The intended audience includes faculty and graduate students in not only computer science, but any field (*e.g.*, geography, epidemiology, wildlife conservation, criminology) where we need to make intelligent inferences from observational data, together with application-specific information. Military planners and analysts, as well as law enforcement officials will also find parts of the book to be of interest.

The authors are indebted to many people who either were involved in some of our research or who facilitated the work in one of many ways.

First, both authors would like to thank their wives—Jana Shakarian and Mary Andrews—for their constant love and support.

Second, the authors would like to thank John Dickerson for serving as a co-author of Chapter 4.

Both authors would also like to thank several university and Army researchers who played a role in the technical development of the work, either merely by listening to the ideas underlying geospatial abduction, or by actively contributing ideas, data, or critiques and comments. In alphabetical order, these include Jeff Brantingham, Damon Earp, Trevon Fuller, Virginia Melissa Holland, Grant Jacoby, Henry Kautz, Sarit Kraus, Steve LaRocca, Dan LaRocque, Scott Lathrop, Roy Lindelauf, Cristian Molinaro, Margo Nagel, Dana Nau, Austin Parker, Jeff Remmel, Anne Rimoin, Patrick Roos, Maria-Luisa Sapino, Brittany Schuetzle, Gerardo Simari, Amy Sliva, Tom Snitch, and Michelle Vanni.

Both authors would like to thank several military personnel who have played important roles in shaping the application of this work to counter-IED operations. These include Charles Otstott, Keith Collyer, Wayne Skill, Geoff Stoker, Tim Warner, and Tim Colson.

Paulo Shakarian would like to thank Eugene Ressler of the U.S. Military Academy for allowing him to pursue the doctoral degree, out of which the geospatial abduction work started. Some of the results described in this book were included as part of Paulo's Ph.D. requirements, which was funded by the West Point Instructor's Program. Paulo would also like to specifically thank Charles Otstott, Tim Warner, and Michelle Vanni for their considerable efforts to help transition some of this work to the U.S. Army.

V.S. Subrahmanian would like to thank Purush Iyer of the Army Research Office in particular for his deep intellectual support of this work and for going to considerable effort to help transition the research to the U.S. Army. Parts of the work reported in this book were generously funded by the Army Research Office under grant W911NF0910206.

West Point, NY,   *Paulo Shakarian*
College Park, MD,   *V.S. Subrahmanian*
                                                                                         August 2011

Contents

Acronyms

#3CNFSAT	Sharp-SAT in 3-Conjunctive Normal Form
#P	Sharp-P
#Pl1Ex3SAT	Planar 1-Exact-3-SAT
#Pl1Ex3MonoSAT	Monotone Pl1Ex3SAT
#Pl3CNFSAT	Planar 3CNFSAT
#Pl3DS	Counting, planar Dominating Set, maximum degree of 3
#PlVC	Sharp Planar Vertex Cover
#SAT	Sharp-SAT
α	alpha
β	beta
AI	Artificial Intelligence
crf	Cutoff Reward Function
CoNP	Complement of Non-deterministic Polynomial Time
DomSet	Dominating Set
exfd	Explanation Function Distribution
FPRAS	Fully Polynomial Randomized Approximation Scheme
FPTAS	Fully Polynomial Time Approximation Scheme
frf	Fall-off Reward Function
GAP	Geospatial Abduction Problem
GCD	Geometric Covering by Discs
GOP	Geospatial Optimization Problem
IED	Improvised Explosive Device
In-#P	Membership in the complexity class #P
In-coNP	Membership in the complexity class coNP
In-NP	Membership in the complexity class NP
ILP	Integer Linear Program
IP	Integer Program
IPB	Intelligence Preparation of the Battlefield
I-REP	Induced Region Explanation Problem
ISW	Institute for the Study of War
k-SEP	k-sized Spatial Explanation Problem

LP	Linear Program
MCA	Maximal Counter-Adversary Strategy
MCA-Exp	Maximal Counter-Adversary Strategy – Explaining
MC	Minimal Cardinality
MCA-LS	Maximal Counter-Adversary Strategy – Local Search
MDP	Markov Decision Process
ME	Maximum Explaining
MILP	Mixed Integer Linear Program
NAI	Named Area of Interest
NP	Non-deterministic Polynomial Time
OAS	Optimal Adversarial Strategy
pdf	Probability Distribution Function
PTIME	Polynomial Time
REP	Region Explanation Problem
rf	Reward Function
SAT	Satisfiability
SC	Set-Cover Problem
SCARE	Spatio-Cultural Abductive Reasoning Engine
SEP	Spatial Explanation Problem
SOMA	Stochastic Opponent Modeling Agents
TD-SEP	Total Distance Spatial Explanation Problem
wrf	Weighted Reward Function
WT-SEP	Weighted Spatial Explanation Problem

Chapter 1
Introduction

Abstract This chapter provides an intuitive, easy to read explanation of what geospatial abduction is. It uses a set of examples to explain what geospatial abduction is, and how it can be used to solve real-world problems in many different domains. Our examples show how geospatial abduction can be used to (i) identify the locations of weapons caches supporting improvised explosive device attacks by terrorists and armed insurgents from information about the locations of the attacks, (ii) identify the possible locations of tigers from information about locations of their kills, (iii) identify habitats that support host animals that carry certain viruses from information about where diseases caused by those viruses occurred, and (iv) identify the location(s) of a burglar from information about where burglaries he carried out occurred. These four examples are used continuously throughout the book to illustrate the mathematical foundations and definitions that are presented in later chapters.

1.1 Motivation

There are numerous applications in the real world in which we observe that certain phenomena occur at various locations and where we wish to infer various "partner" locations that are somehow associated with those observations. Partner locations could be associated with entities that cause the phenomena we observe or facilitate the observations that we observe.

Informally speaking, a *geospatial abduction problem* (GAP) refers to the problem of finding partner locations that best explain a set of observations (at certain locations), in the context of some domain-specific information that tells us something about the relationship between the observations we make and the partner locations that cause, facilitate, support, or are somehow correlated with the observations. Geospatial abduction was first introduced by the authors in [4] and later studied by them in a series of papers [5, 6, 7, 8].

For instance, we have used geospatial abduction to find the locations of the weapons caches that allow insurgents and terrorists in Baghdad, Iraq to carry out improvised explosive device (IED) attacks both on Iraqi civilians, as well as on multinational troops situated there. In this application of the geospatial abduction technique, the observations correspond to the locations of the attacks and the partner locations we wish to find are the locations of the weapons caches that facilitate or support those attacks. Of course, to do so, we must take domain information into account. What kinds of places are suitable locations for weapons caches? Are there operational constraints on the insurgents that somehow constrain how far the weapons caches can be from the locations of the attacks? The answers to these questions constitute *application-specific* information related to the problem of detecting IED caches. A generic algorithm, computational engine, or software tool for geospatial abduction must support application development where the application (in this case, detection of IED weapons caches) developer can explicitly articulate such application-specific information to the GAP engine in a manner that is *uniform* and *application independent*.

As another application, consider the case of tiger conservation. The number of tigers in the world is dwindling rapidly and organizations such as the World Wildlife Fund (WWF) are making heroic efforts to save the tiger. Unfortunately, the tiger is not an easy animal to save. Unlike lions, they are solitary creatures that maintain a very stealthy existence. Their range can be over 100 square miles, often making it difficult to pinpoint exactly where they like to reside at a given time. We have been considering the prospect of identifying relatively small regions where tigers might like to reside based on observations (locations) of tiger kills. Fortunately, after eating its meal, the tiger does not drag away the carcass or skeleton that is left behind, providing researchers and conservationists valuable information on where the tiger has been. In this application as well, we need to take much domain specific information into account (*e.g.*, a wide open space is not a place where it is likely that a tiger will dwell, nor is a place where there is a paucity of prey [10]).

A third application we have worked on is an effort led by epidemiologists at UCLA that involves identifying the habitats of creatures that carry certain viruses. For instance, monkey pox [2] is a deadly disease that kills and/or irreparably damages many children—and even adults—in Africa. It is particularly widespread in the Democratic Republic of Congo. The disease is spread by host animals that are often eaten raw by a hungry, highly malnourished human population, who are desperate for food. Thus, a natural public health question arises. Can we somehow identify the habitats where the host animals live in large numbers so that appropriate public health measures (*e.g.*, extermination of the hosts or other environmentally appropriate actions) can be taken? As in the case of the tigers above, this requires application specific knowledge about the types of environments/habitats that the host animals prefer and/or flourish in.

A fourth application deals with crime. We are all painfully aware of the existence of burglars and home invasions. How can we identify the locations (home or office or even a significant other's house—as long as it is a place where the burglar spends a fair amount of time) of an unknown burglar or home invader by examining the locations where the burglaries or home invasions were committed? In this case, again, domain specific information can be taken into account. For instance, we know that burglaries are usually committed in neighborhoods that the burglar knows, but usually the burglar targets homes that are not too close to either his home or his office or places where he spends a lot of time and is known to others. How do we find the burglar's house or somehow narrow down the space of possible targets?

In the rest of this chapter, we explore these applications in further detail, clearly articulating the issues involved in further detail. In short, this chapter tries to explain what types of real-world problems geospatial abduction is supposed to solve, but not how. Following this chapter, most of the rest of this book will focus on the "how."

1.2 The IED Cache Detection Problem

Improvised Explosive Devices (IEDs) are crude bombs constructed by insurgents to attack an external force. The term IED was first introduced by the British Army in response to attacks by the Irish Republican Army (IRA) in the 1970s. Since then, it has been used by insurgent groups around the world to attack external forces.

Figure 1.1 shows a screenshot of real-world data gleaned from open sources about the locations of IED attacks in Baghdad during the February 2007–November 2008 time frame. The map was generated using the Spatio-Cultural Abductive Reasoning Engine (SCARE) system [4] which in turn used Google Maps to get geographic data. The red push pins show the locations of IED attacks during this time frame.

All the IED attacks shown in Figure 1.1 were believed to have been carried out by Shiite-militia supported by Iran. Experience has shown that these attacks were typically carried out by insurgents who placed their munitions in *weapons caches*. A weapons cache was then used to support one or more attacks.

Of course, the insurgents were not stupid and had no wish to get caught. Weapons caches were chosen carefully. In particular, it was clear that the insurgents could not locate weapons caches within US or international coalition bases. Likewise, they could not locate weapons caches within Sunni neighborhoods of Baghdad because of ongoing ethnic conflict between the Shiites and Sunnis. Last, but not least, we deemed that they could not place weapons caches on the Tigris river because of the probability of being spotted as well as the logistical difficulties involved in transporting munitions from a river to land. [9] contains further work on IED cache placement. The shaded regions in Figure 1.2 shows regions where the IED caches could not be located.

The job of a geospatial analyst is now clear. Is there a way to study the map of Figure 1.1 showing the locations of the IED attacks, together with the map overlays shown in Figure 1.2 showing where caches could not possibly occur, and infer the plausible locations of weapons caches used to support the IED attacks carried out by Shiite insurgents?

Fig. 1.1 SCARE [4] screenshot showing locations of IED attacks in Baghdad during the Feb. 1, 2007 to Nov. 2008 time frame.

The problem is highly non-trivial to solve for several reasons. First, we do not know how many IED caches there are to find. Second, the zones where IED caches cannot be present (as shown in Figure 1.2) are highly irregular in shape—so simple geometric reasoning cannot be a solution. Third, the insurgents are constantly adapting their attack techniques to any counter-measures being taken to find and/or thwart them. Finally, as we shall show in Chapter 2, the problem of finding a set of such cache locations is NP-complete, making it intractable to compute in practice.

We have developed two systems called SCARE [4] and SCARE-S2 [7] that use geospatial abduction. SCARE used a version of geospatial abduction called *point-based geospatial abduction* (studied in Chapter 2) that was applied to the problem of finding IED weapons caches in Baghdad. Using 21 months of data (7 for training, 14 for evaluation), we were able to show that SCARE predicted cache locations that (on average) were within 0.45 miles of the actual locations of caches discovered in Baghdad by coalition forces.

SCARE-S2 was applied to the problem of discovering high value targets (or HVTs) in Helmand and Kandahar provinces of Afghanistan. HVTs were defined to be either depot-level weapons caches (as opposed to smaller caches designated for more immediate use) or insurgent commanders. SCARE-S2 used a different technique than SCARE called *region abduction*, described in detail in Chapter 3, to identify regions in these provinces that were highly likely to contain HVTs. Comparison with real-world data showed that the regions we discovered had a density of HVTs that was 35 times higher than the density of HVTs in the two provinces

Fig. 1.2 SCARE [4] screenshot showing coalition bases, Sunni neighborhoods, and the Tigris River.

considered as a whole. In addition, these regions contained on average 4.8 villages that needed to be searched by US and coalition forces.

1.3 The Tiger Detection Problem

As anybody who has ever gone to a tiger reserve knows, getting to the tiger reserve is easy, but spotting a tiger is hard. Tigers are hunters who live a largely solitary existence and depend on stealth attacks in order to capture prey. At the time this chapter was written, the World Wildlife Fund estimated that there are fewer than 3,200 tigers still living in the wild in the entire world.

Wildlife experts have considerable interest in identifying the precise region where the tigers are living so that appropriate conservation steps can be taken.[1] Consider the Achanakamar Wildlife Sanctuary (AMWLS) in the state of Chattisgarh, India. Tiger conservation experts would like to understand exactly where the tigers reside. In order to do so, the wildlife conservators looking at a map of AMWLS

---

[1] We thank Tom Snitch for suggesting we consider this problem using geospatial abduction techniques after a meeting he had with World Wildlife Fund officials who expressed concern about the need for better tracking of tigers.

need to identify locations in the sanctuary that are feasible for tigers to adopt as their range. This involves a number of issues. For instance:

- The places where the tigers live needs to have a high concentration of *prey* which in the case of this sanctuary includes chital, sambar, as well as wild boar.
- The placess where the tigers live need to have the right kind of vegetation, involving variables such as "canopy cover, canopy height, forest, shrub cover, shrub height" [10, page 563].
- The number of dung pellets found in a given region is also correlated with the suitability of a location for the tiger's habitat as this is closely correlated with the amount of prey in the area (more dung pellets implies more prey).

Thus, wildlife analysts may first plot a "habitat map" showing locations that are suitable for the tiger to live versus locations that are not suitable for the tiger to live, as shown in Figure 1.3 below.

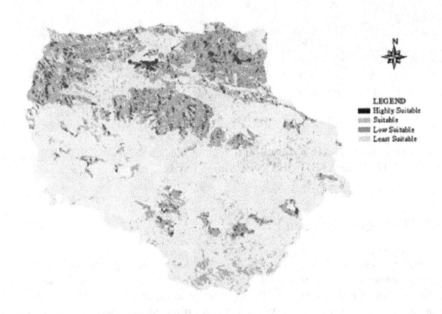

Fig. 1.3 Tiger habitat suitability map for the Achanakamar Wildlife Sanctuary—figure taken from *M. Singh, P.K. Joshi, M. Kumar, P.P. Dash and B.D. Joshi. Development of tiger habitat suitability model using geospatial tools: a case study in Achankmar Wildlife Sanctuary (AMWLS), Chhattisgarh India, Env. Monitoring and Assessment journal, Vol. 155, pages 555-567, 2009.* and reprinted courtesy of Springer.

A wildlife analyst equipped with such a map (stored in the Keyhole Markup Language, or KML, format) can use geospatial abduction through SCARE or SCARE-S2 to upload an Excel file containing information on the location of various tiger kills. Figure 1.4 shows one such example (synthetic data) of locations of tiger kills in AMWLS.

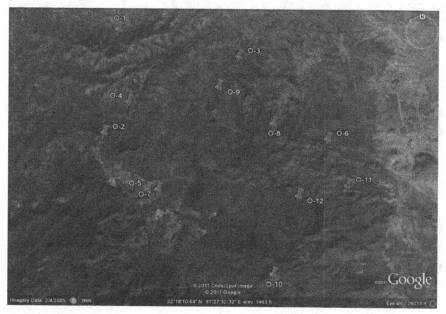

Fig. 1.4 Tiger kill locations in AMWLS. Synthetic data used for example purposes only.

The goal is to now determine where the tiger responsible for the actual kills lives, given both the locations of its kills and the habitat suitability map. Region-based geospatial abduction studied in Chapter 3 provides a suite of techniques to address this problem. Figure 1.5 shows potential locations predicted by SCARE [4].

1.4 The Virus Host Habitat Identification Problem

A related potential application of geospatial abduction, similar to the tiger habitat problem, is that of identifying the habitats of animal hosts that carry certain viruses which cause diseases in human populations.[2] Many such diseases fall into the category of *vector-borne diseases* in which a host transmits a virus to humans, usually via a bite.

Realistic examples of such diseases include diseases spread through mosquito bites (*e.g.*, malaria, chikungunya fever, yellow fever, West Nile encephalitis and other types of encephalitis), diseases spread by rodents and rodent fleas (*e.g.*, plague, monkey pox), diseases caused by ticks and deer flies living on deer (*e.g.*, Lyme dis-

---

[2] We are grateful to Trevon Fuller for thinking of this application.

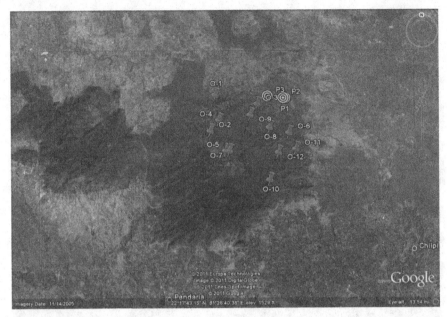

Fig. 1.5 Tiger kill locations with predicted tiger locations in AMWLS. Synthetic data used for example purposes only.

ease, tularemia), diseases caused by various types of flies (*e.g.*, sleeping sickness), and many others.

In such cases, a public health expert might ask himself the question: How can I identify the locations of habitats of hosts (*e.g.*, deer, rodents) that support the organisms (*e.g.*, ticks) that spread these diseases? To do this, the public health expert can use geospatial abduction to carry out the following steps:

- Identify locations where the disease occurred or has been known to occur (perhaps at a certain level of occurrence or higher so that isolated cases do not skew the analysis).
- Identify the properties of habitats (*e.g.*, standing bodies of water in the case of mosquito-borne diseases or the existence of certain types of foliage in the case of deer) that support the host animals.

Based on these two analyses, the public health analyst can easily use a region-based geospatial abduction tool to identify regions which have a high probability of supporting the hosts that carry and spread the disease. Once these regions are identified, appropriate public health actions can be taken, possibly in conjunction with public health authorities.

1.5 The Burglar Detection Problem

Police all over the world are constantly confronted with burglaries. Using a number of forensic techniques, they can often identify which burglaries were committed by the same perpetrator(s). A natural question for criminologists and law enforcement agencies is to figure out how to find the places where the burglar lives or works.

It is well known in criminology [1, 3] that burglars, serial killers, and many other types of criminals often carry out their criminal activities in areas they know well. Typically, this condition of "knowing well" means that either the criminals live in the area where they carry out their crimes, or work there, or grew up there.

Figure 1.6 shows a map of St. Paul, Minnesota, with the locations of various church burglaries explicitly marked via red push pins. This data shows real church burglaries that occurred in 2008–2009, not synthetic information. Moreover, the police in St. Paul believed that these burglaries were all carried out by the same burglar.

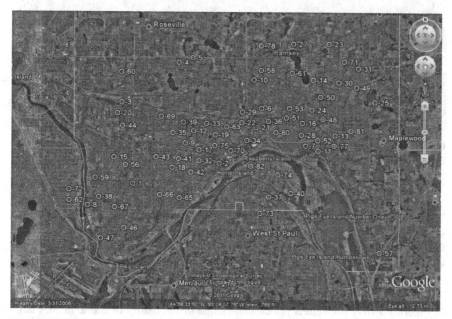

Fig. 1.6 SCARE [4] screenshot showing locations of church burglaries in St. Paul, Minnesota in 2008–2009.

A criminologist or police officer investigating these burglaries might want to give a geospatial abduction system some information. For instance, he might say that he does not believe that a burglar would commit such crimes less than a kilometer from his house or more than seven kilometers from his house (these distances can also be automatically learned from historical data or explicitly provided by an expert). In addition, he might mark certain regions on the map as unlikely places for the burglar

to have his home or office. Such *excluded* regions are shown in Figure 1.7. Note
that in this example, these are only "notional" excluded regions and real excluded
regions would need to be inserted by a domain expert (*e.g.*, a St. Paul, MN, police
officer investigating the burglaries).

Fig. 1.7 SCARE [4] screenshot showing regions in St. Paul, Minnesota, that were excluded as
potential locations for the church burglar.

Last, but not least, we would like our geospatial abduction system to generate
"predicted" locations for the church burglar. It is too hard to designate whether these
predicted locations represent his home or his office—rather, they represent locations
that are most likely to be locations where he has a significant presence. Figure 1.8
shows the St. Paul, Minnesota, map, together with yellow bull's-eyes reflecting predicted locations. Again, we emphasize that these are *notional* predicted locations;
even though the church burglary data we use is real data, our exclusion zones shown
in Figure 1.7 may not reflect police knowledge of the reality of crime in St. Paul,
and hence, the results shown in Figure 1.8 may be incorrect. Our purpose in this
example is to show how such a system should work.

1.6 Other Applications

The preceding sections highlight four real-world applications in which geospatial
abduction is currently or could be employed. However, the space of possible ap-

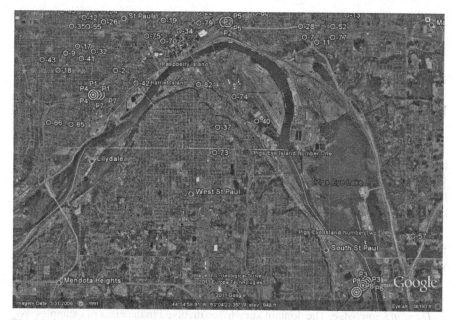

Fig. 1.8 SCARE [4] screenshot showing locations of predicted locations of the burglar with respect to the church burglaries in St. Paul, Minnesota, in 2008–2009.

plications for geospatial abduction is really much larger; we highlight a few more examples here, though we will be unable to consider them in further detail in the rest of the book.

One important application area deals with environmental pollutants in a body of water. Often times, water is contaminated by unscrupulous organizations or companies that dump toxic waste into a body of water. We do not always know who the responsible party is, but identifying the location(s) where the dumping is likely to be occurring allows environmental authorities and police to target their surveillance efforts with a view to catching the culprits. In this case, the observations are the locations where the pollution was discovered (*e.g.*, contaminated water), and the partners we want to find are the locations where the polluting substances are introduced into the water or into the ground. Domain information specifies how the contamination spreads—either through the water or through the ground.

The same principle also applies to pollution in the ground: we see contaminants at various locations on the ground and we would like to infer the source of these contaminants. The source may be a leak in a network of pipes distributing the substance that is leaking, or an explicit attempt to dump pollutants, or simply an accident. Knowing the location from which the pollutant is coming can play a key role in helping solve the problem.

Another important application is identifying the location of illegal drug labs or distribution centers from information about the locations where various drug dealers were arrested. Alternatively, with aerial surveillance of the coca plant in countries like Colombia and Peru, we know the locations of the base crop that is converted into an illegal substance. Based on the locations of these fields, can we infer the locations of the labs that convert these crops into illegal drugs?

1.7 Conclusion

We see in this chapter that geospatial abduction in different forms can be used to help address a wide variety of problems that have a significant geospatial character. We have only described a small number of problems that geospatial abduction can help with. As the technique is studied more extensively, we believe there will be far more applications.

All of these geospatial abduction applications described have the following characteristics:

1. *Observations*. There is a set of observations that we start with. The set of observations could be the locations where IED attacks occurred, where disease outbreaks occurred, where tiger kills were observed, or where pollutants were spotted.
2. *Domain knowledge*. The domain knowledge involved in the class of examples we have discussed include two types of phenomena.

   a. Information about the distances between the locations or regions we are trying to find (*e.g.*, locations of IED weapons caches or regions where the tiger responsible for certain kills may be) and the observations that are causally linked to the observation; and
   b. Maps showing which locations or regions on the ground satisfy various "feasibility requirements" (*e.g.*, having the appropriate type and quantity of prey in the case of the tiger habitat identification problem, or having the right kinds of populations for insurgents to blend into after carrying out IED attacks).

These inputs can be specified in a variety of ways; however, in later chapters of this book, we will show that these inputs can often be specified in a highly (syntactically) restricted format that makes them easy to manipulate computationally.

Once these inputs are provided, we will show in the rest of this book, how we can find a set of places that best explains the observations in our application while being consistent with the provided domain knowledge.

References

1. Brantingham, P., Brantingham, P. 2008. Crime Pattern Theory. In Enviromental Criminology and Crime Analysis, R. Wortley and L. Mazerolle, Eds., pp. 78–93.
2. Rimoin, A. *et al*. Endemic Human Monkeypox, Democratic Republic of Congo, 2001–2004, Emerging Infectious Diseases, 13, 6, pp. 934–937, 2007.
3. Rossmo, D. K., Rombouts, S. 2008. Geographic Profiling. In Enviromental Criminology and Crime Analysis, R. Wortley and L. Mazerolle, Eds. pages 136–149.

4. Shakarian, P., Subrahmanian, V.S., Sapino, M.L. SCARE: A Case Study with Baghdad, Proc. 2009 Intl. Conf. on Computational Cultural Dynamics (eds. D. Nau, A. Mannes), Dec. 2009, AAAI Press.
5. Shakarian, P., Subrahmanian, V.S., Sapino, M.L. 2012. GAPS: Geospatial Abduction Problems, ACM Transactions on Intelligent Systems and Technology (TIST), 3, 1, to appear.
6. Shakarian, P., Subrahmanian, V.S. Region-based Geospatial Abduction with Counter-IED Applications, accepted for publication in: Wiil, U.K. (ed.).Counterterrorism and Open Source Intelligence, Springer Verlag Lecture Notes on Social Networks, to appear, 2011.
7. Shakarian, P., Nagel, M., Schuetzle, B., Subrahmanian, V.S. 2011. Abductive Inference for Combat: Using SCARE-S2 to Find High-Value Targets in Afghanistan, in Proc. 2011 Intl. Conf. on Innovative Applications of Artificial Intelligence, Aug. 2011, AAAI Press.
8. Shakarian, P., Dickerson, J., Subrahmanian, V.S. 2012. Adversarial Geospatial Abduction Problems, ACM Transactions on Intelligent Systems and Technology (TIST), to appear.
9. Shakarian, P., Otstott, C. What is Old is New: Countering IEDs by Disrupting the Weapon Supply, Military Review, pp. 46–52, 2011.
10. Singh, M., Joshi, P.K., Kumar, M., Dash, P.P., Joshi, B.D. Development of tiger habitat suitability model using geospatial tools—a case study in Achankmar Wildlife Sanctuary (AMWLS), Chhattisgarh India, Env. Monitoring and Assessment journal, Vol. 155, pp. 555–567, 2009.

Chapter 2
Point-based Geospatial Abduction

Abstract In this chapter, we will define geospatial abduction problems (GAPs) formally. In particular, we define a GAP to consist of several parts—a set of observations, two non-negative real numbers α, β denoting lower and upper distance bounds respectively between locations we want to discover (*e.g.*, IED weapons cache locations) and observations, and a "feasibility" predicate. Based on this, the goal of a *point-based* GAP is to find a set of *points* that jointly fall within the distance bounds and are feasible according to the feasibility predicate. We provide results both on the complexity of GAPs as well as detailed algorithms to solve point-based GAPs efficiently in practice. We conclude the chapter with experimental results on real-world IED cache discovery in Baghdad showing that our algorithms work very well in practice.

2.1 Introduction

Chapter 1 provides a large set of examples of geospatial abduction problems drawn from many real-world scenarios including defense and security applications, epidemiology and public health applications, pollution and environmental applications, wildlife preservation and conservation applications, and criminology and law enforcement applications. In this chapter, we will develop the formal mathematics required to efficiently solve geospatial abduction problems.

We first need to formally define a geospatial abduction problem in such a way that the applications described in Chapter 1 can be captured as special cases. Figure 2.1 shows the proposed components of our GAP definition. From this figure, we see that a GAP consists of three major components:

Observations. A set of observations describing locations corresponding to the phenomenon under study (*e.g.*, IED attack locations, locations of tiger kills, locations where a disease was observed, locations where a burglary occurred, locations where pollution was detected).

Distance Constraints.   A pair of real numbers α, β describing a lower bound (α)
and an upper bound (β) on the distances between an observation and a related
location (or *partner* location) we would like to discover. Partner locations we
might like to discover in our applications include the locations of IED weapons
caches, the locations where the tiger likes to reside, the regions where a disease
host flourishes, the home and/or office locations of the burglar, and so forth.

Feasibility Predicate.   A feasibility predicate allows an application developer to
specify which points on the map are (or are not) potential locations for a partner.
For instance, a feasibility predicate in the IED detection problem specifies which
points satisfy application-dependent criteria that an IED cache location should
satisfy. A defense analyst may articulate this, for instance, by saying that such
locations cannot lie within a coalition base or within an area occupied by an
ethnic group opposed strongly to the group carrying out the attacks. In our tiger
application, the feasibility predicate might rule out areas with no forest canopy
and/or very little dense vegetation.

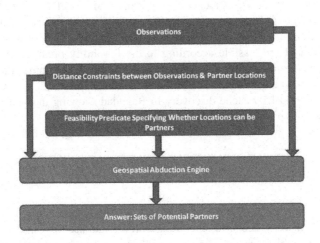

Fig. 2.1 Inputs and components of a canonical geospatial abduction definition..

Based on these inputs, the goal of a geospatial abduction system is to find a set
of "partner" locations that:

1. Explain all the observations we see.
2. Fall within the (α, β) distance constraints.
3. Are all feasible according to the feasibility predicate that has been chosen for the
   application.

*It is important to note that the observations, the α, β numbers, and the feasibility
predicate are all inputs to a GAP solver and can be tuned by the analyst or user*

to meet his needs and reflect his own expertise. Moreover, we will show that if historical data is available, we can easily learn the α, β numbers automatically from the historical data using a simple algorithm.

The rest of this chapter is organized as follows. We first present a formal definition of geospatial abduction problems, together with several variants. We then study the computational complexity of these problems, followed by a presentation of algorithms—both exact and heuristic—to solve arbitrary GAPs. We conclude with a set of experimental results we have derived on GAPs, showing both that the algorithms work efficiently in practice and that they are highly accurate.

2.2 Point-based GAPs, Formalized

With the exception of Chapter 3, we assume throughout this book that a map (resp. space) is represented by a two dimensional grid of size $M \times N$ where $M, N \geq 1$ are integers. The space of all points, therefore, is the set $\mathscr{S} = \{i \mid 0 \leq i < M\} \times \{j \mid 0 \leq j < N\}$.

Each point $(i, j) \in \mathscr{S}$ represents the unit square on the map whose lower left corner is the point (i, j). Each such unit cell has closed left and lower edges, and open edges on the right and top. This is consistent with virtually all geospatial data structures such as various types of quadtrees and R-trees [24] and is also consistent with how most of the major geographic information systems represent spatial data.

The developer of a geospatial abduction application can choose M and N to be as large as he or she wants. By choosing large M, N to represent a particular region on the ground, each cell in the grid represents a smaller region on the ground. Thus, increased M, N yields increased map resolution. As the GAP application developer gets to choose M, N, he or she can effectively pick the appropriate resolution for a specific application.

We assume that all observations occur within space $\mathscr{S}$. We use the artificial space shown in Figure 2.2 throughout this chapter to illustrate the concepts we introduce. Throughout this book, we assume that $\mathscr{S}$ has an associated *distance function d* satisfying the usual properties of such distance functions. Standard distance functions d in topology and metric spaces satisfy three simple axioms defined below.

Definition 2.1. A distance function d on $\mathscr{S}$ is a mapping from $\mathscr{S} \times \mathscr{S}$ to the reals such that:

- $d(x, x) = 0$. This axiom says that the distance between a point and itself is 0.
- $d(x, y) = d(y, x)$. This axiom says that the distance between two points is not dependent on the order in which the two points occur when the distance function is invoked, *i.e.*, the distance function is symmetric.
- $d(x, y) + d(y, z) \geq d(x, z)$. This inequality says that the distance function satisfies the triangle inequality.

There are numerous distance functions that satisfy these axioms. Examples include:

- *Euclidean distance* where $d_e((x_1, y_1), (x_2, y_2)) = \sqrt{(x_1 - x_2)^2 + (y_1 - y_2)^2}$.
- *Manhattan distance* where $d_m((x_1, y_1), (x_2, y_2)) = |x_1 - x_2| + |y_1 - y_2|$.
- *Road distance* $d_r((x_1, y_1), (x_2, y_2))$ is defined as the length of the shortest path (along roads) between two points (x_1, y_1) and (x_2, y_2), assuming the existence of a road network.

The methods used in this chapter apply to *any notion of distance between two points as long as the three distance axioms described above are satisfied.*

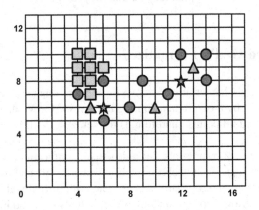

Fig. 2.2 An example space. Red dots denote observations. Yellow squares denote infeasible locations. Green stars show one (0,3) explanation, while pink triangles show another (0,3) explanation.

Definition 2.2 (observation). An *observation* $\mathcal{O}$ is any finite subset of $\mathcal{S}$.

Consider the geospatial universe shown in Figure 2.2. Let us see how the observations in this "toy" figure apply to our real-world applications.

1. In our IED Detection application, these observations correspond to locations where IED attacks occurred.
2. In our Tiger Detection application, these observations correspond to locations where tiger kills were spotted by wildlife conservationists.
3. In our Virus Host Habitat Identification application, these observations correspond to locations where a disease outbreak (caused by the virus in question) was detected.
4. In our Burglar Detection problem, these observations correspond to the locations where the burglar committed his burglaries.

As we have mentioned in Chapter 1, there are serious constraints on where the locations we are searching for (*e.g.*, the location of IED weapons caches or the locations where the tiger resides or the environments where the virus' host flourishes or the places where the burglar's house could be) can be. For instance, we might eliminate a US military base as a potential location for an enemy's IED weapons cache as such areas are usually well secured by US military personnel. Likewise, we can eliminate areas with no ground cover and/or or no prey—tigers are masters of concealment and are unlikely to want to stay in open areas for any significant length of time. In our virus host example, certain areas may be inhospitable to certain hosts, depending upon the biology of the host animal. In our burglar detection problem, we can probably exclude police stations and military bases from the space of locations where the burglar might live and/or work, as personnel who work in these areas are subject to stringent background checks.

The concept of a *feasibility predicate* defined below captures which points in $\mathscr{S}$ are feasible, and which ones are not.

Definition 2.3 (feasibility predicate). A feasibility predicate feas is a function from $\mathscr{S}$ to $\{\mathsf{TRUE}, \mathsf{FALSE}\}$.

Thus, $\mathsf{feas}(p) = \mathsf{TRUE}$ means that point p is feasible and must be considered in the search. Figure 2.2, denotes infeasible places via a yellow square. Throughout this chapter, we assume that feas is an arbitrary, but fixed predicate.[1] Further, as feas is defined as a function over $\{\mathsf{TRUE}, \mathsf{FALSE}\}$, it can allow for user input based on analytical processes currently in place.

For instance, in the military, analysts often create Modified Combined Obstacles Overlays (MCOOs for short) where "restricted terrain" is deemed infeasible [31]. We can also easily express feasibility predicates in a Prolog-style language—we can easily state (in the burglar identification example) that point p is considered feasible if p is within R units of distance from some observation and p is not in the water. Likewise, in the case of the tiger identification example, we could say that a point p is feasible if p is within R_1 units of places with dense land-cover and where the amount of scat (associated with prey) exceeds a certain amount. A Prolog-style language that can express such notions of feasibility is the hybrid knowledge base paradigm [17] in which Prolog-style rules can directly invoke a GIS system.

Suppose now that α, β are two numbers with $0 \leq \alpha \leq \beta$.

1. In our IED cache detection application, for example, α, β say that for every observation (*i.e.*, IED attack), there must be an IED cache which is at least α units from the observation and at most β units from the observation.
2. In our tiger detection application, the α, β numbers say that for every observation (*i.e.*, tiger kill location), there must be a tiger preferred residence which is at least α units from the observation and at most β units from the observation. In a sense, this describes the tiger's territory or range.
3. In our virus host habitat detection problem, the α, β numbers say that every location where a disease caused by the virus is found must have a corresponding host habitat that is at least α units from the disease location and at most β units of distance from the disease location.

---

[1] We also assume throughout the chapter that feas is computable in constant time. This is a realistic assumption, as for most applications, we assume feas to be user-defined. Hence, we can leverage a data structure indexed with the coordinates of $\mathscr{S}$ to allow for constant-time computation.

4. In our burglar detection problem, as in the case of the tiger detection problem, the
   α, β numbers say that for every observation (*i.e.*, burglary location), there must
   be a burglar "hangout" (such as his house or place of work) which is at least α
   units from the observation and at most β units from the observation. In a sense,
   this describes the burglar's home turf—it is just not too close to his home/office.

We now come to the important definition of an *explanation*. Intuitively, given a
set $\mathcal{O}$ of observations, an explanation is a set of points $\mathcal{E}$ such that every point in
$\mathcal{E}$ is feasible and such that for every observation, there is a point in the explanation
that is at least α units away from the observation, but no more than β units away
from the observation.

Definition 2.4 ((α, β) explanation). Suppose $\mathcal{O}$ is a finite set of observations, $\mathcal{E}$ is
a finite set of points in $\mathcal{S}$, and $0 \le \alpha < \beta \le 1$ are some real numbers. $\mathcal{E}$ is said to
be an (α, β) *explanation* of $\mathcal{O}$ iff:

- $p \in \mathcal{E}$ implies that feas$(p) =$ TRUE, *i.e.*, all points in $\mathcal{E}$ are feasible, and
- $(\forall o \in \mathcal{O})(\exists p \in \mathcal{E}) \, \alpha \le d(p,o) \le \beta$, *i.e.*, every observation is neither too close
  nor too far from some point in $\mathcal{E}$.

Thus, an (α, β) explanation is a set of points. Each point must be feasible and every
observation must have an analogous point in the explanation which is neither too
close nor too far.

For instance, consider our tiger detection problem. Suppose we found tiger kills
at various locations (*i.e.*, at locations in the set $\mathcal{O}$) and suppose further that we know
all these kills were carried out by the same tiger (wildlife experts can identify tiger
pug marks with the same precision with which they can identify human fingerprints).
Wildlife experts have extensively studied how far tigers travel while seeking prey.
They can place bounds (corresponding to the α, β values) on these prey-seeking
travel distances. In the same way, the feasibility predicate specifies the locations
that constitute a suitable habitat for tigers (*e.g.*, lots of ground cover, lots of prey).
Thus, an explanation $\mathcal{E}$ is a set of points that are all feasible and such that for every
tiger kill p_1 attributed to this specific tiger, there is a location p_2 in $\mathcal{E}$ whose distance
from p_1 is in the interval $[\alpha, \beta]$.

The same phenomenon is true of burglars and serial killers. Criminologists [2, 23]
have observed that various types of criminals also tend to carry out their criminal
activities in areas they know well, but not too close to their home bases. This makes
sense—the criminals know the region well, know good targets for their nefarious activities, and have knowledge that allows them to get away quickly after their crime.
This accounts for their not straying too far from their home base (*e.g.*, home, office,
significant other's house, etc) and thus explains the upper bound, β. At the same
time, the criminals may not want to carry out their activities right next to their home
base; in the case of a murder, for instance, police usually question neighbors and
investigate certain people living within a certain distance of the murder location.
The criminal may not want to come to the police's attention in this way by staging
his crimes too close to his home base.

Thus, we see that an (α, β) explanation $\mathcal{E}$ for a given set of observation $\mathcal{O}$ must appropriately explain each observation $o \in \mathcal{O}$ by postulating the existence of a partner location $p \in \mathcal{E}$ such that

- The distance $d(p, o)$ between p and o lies within the $[\alpha, \beta]$ interval; and
- p is a feasible location.

Of course, given an (α, β) explanation $\mathcal{E}$, there may be an observation $o \in \mathcal{O}$ such that there are two (or more) points $p_1, p_2 \in \mathcal{E}$ satisfying the conditions of the second bullet of Definition 2.4 above. If $\mathcal{E}$ is an explanation for $\mathcal{O}$, a *partnering function* $\wp_{\mathcal{E}}$ is a function from $\mathcal{O}$ to $\mathcal{E}$ such that for all $o \in \mathcal{O}$, $\alpha \leq d(\wp_{\mathcal{E}}(o), o) \leq \beta$. $\wp_{\mathcal{E}}(o)$ is said to be o's *partner* according to the partnering function $\wp_{\mathcal{E}}$. We now present a simple example of (α, β) explanations.

Example 2.1. Consider the observations in Figure 2.2 and suppose $\alpha = 0, \beta = 3$. Then the two green stars denote an (α, β) explanation, *i.e.*, the set $\{(6,6), (12,8)\}$ is a $(0,3)$ explanation. So is the set of three pink triangles, *i.e.*, the set $\{(5,6), (10,6), (13,9)\}$ is also an $(0,3)$ explanation.

The basic problem that we wish to solve in this chapter is the following.

The Simple (α, β) Explanation Problem (SEP).
INPUT: Space $\mathcal{S}$, a set $\mathcal{O}$ of observations, a feasibility predicate feas, and numbers $0 \leq \alpha < \beta \leq 1$.
OUTPUT: "Yes" if there exists an (α, β) explanation for $\mathcal{O}$, and "no" otherwise.

A variant of this problem is the k-**SEP** problem which requires, in addition, that $\mathcal{E}$ contains k elements or less, for $k < |\mathcal{O}|$. Yet another variant of the problem tries to find an explanation $\mathcal{E}$ that is "best" according to some cost function.

Definition 2.5 (cost function χ). A cost function χ is a mapping from explanations to non-negative reals.

We will assume that cost functions are designed so that the smaller the value they return, the more desirable an explanation is. Intuitively, a cost function allows us to assess how "good" an explanation function is.

Some example cost functions are given below. The simple one below merely looks at the mean distances between observations and their partners.

Example 2.2 (Mean-distance). Suppose $\mathcal{S}, \mathcal{O}, \text{feas}, \alpha, \beta$ are all given and suppose $\mathcal{E}$ is an (α, β) explanation for $\mathcal{O}$ and $\wp_{\mathcal{E}}$ is a partnering function. We could initially set the cost of an explanation $\mathcal{E}$ (with respect to this partnering function) to be:

$$\chi_{\wp_{\mathcal{E}}}(\mathcal{E}) = \frac{\Sigma_{o \in \mathcal{O}} d(o, \wp_{\mathcal{E}}(o))}{|\mathcal{O}|}.$$

Suppose $ptn(\mathcal{E})$ is the set of all partner functions for $\mathcal{E}$ in the above setting. Then we can set the cost of $\mathcal{E}$ as:

$$\chi_{mean}(\mathcal{E}) = \inf\{\chi_{\wp_{\mathcal{E}}}(\mathcal{E}) \mid \wp_{\mathcal{E}} \in ptn(\mathcal{E})\}.$$

The above definition removes reliance on a single partnering function as there may be several partnering functions associated with a single explanation. We illustrate this definition using a highly simplified version of our tiger kill example.

Example 2.3. Suppose wildlife experts have found tiger kills at the locations shown in the space $\mathcal{S}$ depicted in Figure 2.3. By analyzing the tiger's pug marks, they know these kills are all those of the same tiger. Points $\{o_1, o_2, o_3\}$ indicate locations of evidence of the tiger kills—they jointly constitute the set $\mathcal{O}$. Points $\{p_1, p_2, \ldots, p_8\}$ indicate feasible residence locations for the tiger. The concentric rings around each element of $\mathcal{O}$ indicate the distance $\alpha = 1.7$km and $\beta = 3.7$km. These might denote the fact that the tiger usually cannot catch prey near its residence (perhaps because the prey are aware of the tiger's hangout) and the fact that the tiger does not want to go too far away for its prey.

The set $\{p_3, p_6\}$ is a valid $(1.7, 3.7)$ explanation for the set $\mathcal{O}$ of observations. However, we note that observation o_2 can be partnered with both p_3 and p_6.

If we wish to partner observations with the nearest partner, we see that $d(o_2, p_3) = 3$km and $d(o_2, p_6) = 3.6$km, hence p_3 is the partner for o_2 such that the distance is minimized.

It is easy to see that there are many possible explanations for a given set of observations. For instance, according to the definition of an explanation, any super-set $\mathcal{E}'$ of an explanation $\mathcal{E}$ is an explanation as long as the points in $\mathcal{E}' - \mathcal{E}$ are feasible. In fact, members of $\mathcal{E}' - \mathcal{E}$ need not even be within the distance bounds α, β from any observation—they could be completely extraneous. For instance, the reader can easily verify that the set $\{p_3, p_6\}$ of Example 2.3 stays an explanation irrespective of what other points we add to the set $\{p_3, p_6\}$. In fact, any superset of $\{p_3, p_6\}$ consisting only of feasible points is a valid solution to **SEP**—and as long as the size of that super set is k or less, it is also a valid solution to k-SEP.

Thus, we need ways of comparing explanations against each other to determine whether one explanation is better than another. We do this through the concept an "optimal" explanation—an explanation that minimizes cost. Cost is represented by a *cost function* χ that assigns a non-negative real number to each explanation. Intuitively, the higher the cost of an explanation, the less desirable it is.

Definition 2.6. Suppose $\mathcal{O}$ is a finite set of observations, $\mathcal{E}$ is a finite set of points in $\mathcal{S}$, $0 \leq \alpha < \beta \leq 1$ are some real numbers, and χ is a cost function. $\mathcal{E}$ is said to be an *optimal* (α, β) explanation iff $\mathcal{E}$ is an (α, β) explanation for $\mathcal{O}$ and there is no other (α, β) explanation $\mathcal{E}'$ for $\mathcal{O}$ such that $\chi(\mathcal{E}') < \chi(\mathcal{E})$.

We present an example of optimal (α, β) explanations below.

Example 2.4. Consider the tiger from Example 2.3 whose behavior is depicted in Figure 2.3 (above). While $\{p_3, p_6\}$ is a valid solution for the k-**SEP** problem ($k = 2$), it does not optimize mean distance. In this case, the mean distance would be 3km.

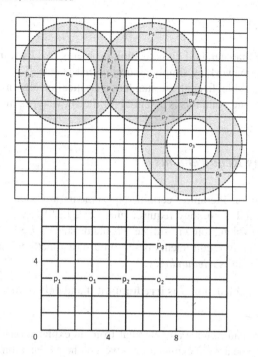

Fig. 2.3 Above: Points $\{o_1, o_2, o_3\}$ indicate locations of tiger kills (*i.e.*, the set $\mathcal{O}$ of observations). Points $\{p_1, p_2, \ldots, p_8\}$ indicate feasible dwellings for the animal. The concentric rings around each element of $\mathcal{O}$ indicate the distance $\alpha = 1.7km$ and $\beta = 3.7km$. **Below:** Points $\{p_1, p_2, p_3\}$ are feasible for crime scenes $\{o_1, o_2\}$. $\{p_1, p_2\}$ are safe houses within a distance of $[1, 2]$ km. from crime scene o_1 and $\{p_2, p_3\}$ are safe-houses within a distance of $[1, 2]$ km. from crime scene o_2.

However, the solution $\{p_3, p_7\}$ provides a mean-distance of 2.8km and hence, this solution would be a better explanation.

Suppose we now consider a burglar who has struck at locations $\mathcal{O} = \{o_1, o_2\}$ and we want to locate his "safe houses" (*e.g.*, his home or his office or his significant other's home). The points $\{p_1, p_2, p_3\}$ are feasible locations. This is depicted in Figure 2.3 (below). Based on historical data, suppose we know that burglars strike at locations that are at least 1km away from a safe house and at most 2km from the safe house ($\alpha = 1$, $\beta = 2$). Thus, for $k = 2$, any valid explanation of size 2 provides an optimal solution with respect to mean-distance as every feasible location for a safe house is within 2km of a crime scene.

We are now ready to define the cost-based explanation problem.

The Cost-based (α, β) Explanation Problem.
INPUT: Space $\mathscr{S}$, a set $\mathcal{O}$ of observations, a feasibility predicate feas, numbers $0 \leq \alpha < \beta \leq 1$, a cost function χ and a real number $v > 0$.
OUTPUT: "Yes" if there exists an (α, β) explanation $\mathscr{E}$ for $\mathcal{O}$ such that $\chi(\mathscr{E}) \leq v$,

and "no" otherwise.

It is easy to see that standard classification problems like k-means[2] can be captured within our framework by simply assuming that $\alpha = 0$, $\beta > \max(M, N)^2$ and that all points are feasible. In contrast, standard classification algorithms cannot take feasibility into account—and this is essential for the above types of applications.

2.2.1 Parsimony Requirements

Typically in abduction literature, certain requirements are added to the explanation. These are called parsimony requirements [21]. In SEP, we enforce no such requirement—any explanation will suffice. Problems such as k-**SEP** and cost-based SEP enforce parsimony requirements based on a cost function. Another common parsimony requirement is *irredundancy*.

Definition 2.7. An explanation $\mathscr{E}$ is **irredundant** if and only if no strict subset of $\mathscr{E}$ is an explanation.

Intuitively, if we can remove any element from an explanation—and this action causes it to cease to be a valid explanation—we say the explanation is irredundant.

Example 2.5. Figure 2.4 shows a map of burglaries depicted in a 18×14 grid. The distance between grid squares is 100 meters. Observation set $\mathscr{O} = \{o_1, o_2, o_3, o_4, o_5\}$ represents the location of the individual burglaries. Based on an informant or from historical data, law enforcement officials know that the burglar frequently stays within 150–320 meters from the center mass of each incident (*i.e.*, in a geospatial abduction problem, we can set $[\alpha, \beta] = [150, 320]$). Furthermore, based on the terrain, the law enforcement officials are able to discount certain areas (shown in black on Figure 2.4, a feasibility predicate can easily be set up accordingly). Based on Figure 2.4, the set $\{p_{40}, p_{46}\}$ is an irredundant explanation. The sets $\{p_{42}, p_{45}, p_{48}\}$ and $\{p_{40}, p_{45}\}$ are also irredundant explanations.

Although irredundancy is a good way to apply Occam's Razor to a geospatial abduction problem, there still exist a potentially exponential number of such solutions. Hence, an algorithm that simply returns an irredundant solution could potentially produce results with a high degree of non-determinism. Hence, in this chapter we will often attempt to find explanations of *minimal cardinality*, which is the natural optimization problem associated with k-**SEP**. We will refer to this problem—that of finding an (α, β) explanation for a given set of observations that is of minimal cardinality—as MINSEP.

---

[2] See [1] for a survey on classification work.

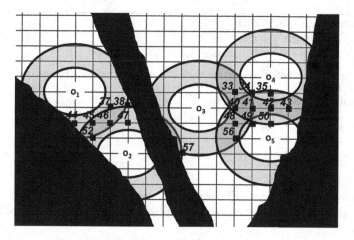

Fig. 2.4 Map of burglaries for Example 2.5. For each labeled point p_i, the "p" is omitted for readability.

Algorithm 1 (STRAIGHTFORWARD-SEP)

INPUT: Space $\mathscr{S}$, a set $\mathscr{O}$ of observations, a feasibility predicate feas, real numbers $\alpha \geq 0, \beta > 0$
OUTPUT: Set $\mathscr{E} \subseteq \mathscr{S}$ of size k (or less) that explains $\mathscr{O}$

1. For each $o \in \mathscr{O}$, let $P_o = \{p \in \mathscr{S} \mid feas(p) = \text{TRUE} \wedge \alpha \leq d(p,o) \leq \beta\}$. Thus, P_o is the set of of locations that are feasible and that are within the α, β distance bounds from o.
2. If it is the case that $P_o \neq \emptyset$ for all o, then we return "yes". Otherwise we return "no."

2.3 Complexity of GAP Problems

In this section, we study the computational complexity of geospatial abduction problems.

We start by observing that the Simple Explanation Problem (**SEP**) can be easily solved in PTIME. To see why, consider the STRAIGHTFORWARD-SEP algorithm.

It is easy to see that this algorithm correctly solves **SEP**, but does so by creating an explanation that is potentially redundant (*e.g.*, by taking the union of all the sets P_o or by taking any member of the cross product of all the sets P_o.)

An alternative naive algorithm would find the set F of all feasible points and return "yes" if and only if for every observation o, there is at least one point $p \in F$ such that $\alpha \leq d(p,o) \leq \beta$. In this case, F is the explanation produced—but it is a very poor explanation.

In the burglar example, F merely tells the police to search all feasible locations without trying to do anything intelligent. Likewise, in the case of the tiger detection problem, it tells wildlife conservationists that the tiger could be in any feasible location.

Algorithm 2 (GCD-TO-KSEP)

INPUT: Instance of GCD $\langle S, P, b, k \rangle$
OUTPUT: Instance of k-**SEP** $\langle \mathscr{S}, \mathscr{O}, \text{feas}, \alpha, \beta, k' \rangle$

1. Set $\mathscr{S}$ to be a set of lattice points in the Euclidean plane that include all points in P
2. Set $\mathscr{O} = P$
3. Let $feas(x) = \text{TRUE}$ iff $x \in P$
4. Set $\alpha = 0$
5. Set $\beta = b/2$
6. Set $k' = k$

In contrast to SEP, the k-SEP problem allows the user to constrain the size of the explanation so that "short and sweet" explanations that are truly meaningful are produced.

The following result states that k-**SEP** is NP-Complete. The proof of NP-hardness is via a reduction from the *Geometric Covering by Discs* (GCD) [14] that is known to be NP-complete.

Theorem 2.1. *k-SEP is NP-Complete.*

Proof. **Geometric Covering by Discs.** (GCD)
INPUT: A set P of integer-coordinate points in a Euclidean plane, positive integers $b > 0$ and $k < |P|$.
OUTPUT: "Yes" if there exist k discs of diameter b centered on points in P such that there is a disc covering each point in P, and "no" otherwise.
CLAIM 1: k-**SEP** is in the complexity class NP.
Suppose a non-deterministic algorithm can guess a set $\mathscr{E}$ that is a k-sized simple (α, β) explanation for $\mathscr{O}$. We can check the feasibility of every element in $\mathscr{E}$ in $O(|\mathscr{E}|)$ time and compare every element of $\mathscr{E}$ to every element of $\mathscr{O}$ in $O(|\mathscr{O}|^2)$ time. Hence, k-**SEP** is in the complexity class NP as we can check the solution in polynomial time.

CLAIM 2: k-**SEP** is NP-Hard.
We use the polynomial algorithm GCD-TO-KSEP to take an instance of GCD and create an instance of k-**SEP**.

CLAIM 2.1: If there is a k'-sized simple (α, β) explanation for $\mathscr{O}$, then there are k discs, each centered on a point in P of diameter b that cover all points in P.
Let $\mathscr{E}$ be the k'-sized simple (α, β) explanation for $\mathscr{O}$. Suppose by way of contradiction, that there are not k discs, each centered on a point in P of diameter b that cover all points in P. As $k' = k$, and all elements of $\mathscr{E}$ must be in P by the definition of feas, let us consider the k discs of diameter b centered on each element of $\mathscr{E}$. So, for these discs to not cover all elements of P, there must exist an element of P, that is not covered by a disc. As $P = \mathscr{O}$, then there must exist an element of $\mathscr{O}$ outside of one of the discs. Note that all elements of $\mathscr{O}$ are within a distance β of an element of $\mathscr{E}$ by the definition of a k'-sized simple (α, β) explanation (as $\alpha = 0$). As $\beta = b/2$,

each element of $\mathcal{O}$ falls inside a disc of diameter b centered on an element of $\mathcal{E}$, thus falling within a disc and we have a contradiction.

CLAIM 2.2: If there are k discs, each centered on a point in P of diameter b that cover all points in P then there is a k'-sized simple (α, β) explanation for $\mathcal{O}$.
Let set E be the set of points that are centers of the k discs. We note that $E \subseteq P$. Assume by way of contradiction, that there is no k'-sized simple (α, β) explanation for $\mathcal{O}$. Let us consider if E is a k'-sized simple (α, β) explanation for $\mathcal{O}$. As $k = k'$, $\alpha = 0$, and all points of E are feasible, there must be some $o \in \mathcal{O}$ such that $\forall e \in E$, $d(e, o) > \beta$. As $\mathcal{O} = P$, we know that all points in $\mathcal{O}$ fall in a disc centered on a point in E, hence each $o \in \mathcal{O}$ must be a distance of $b/2$ or less from a point in E. As $\beta = b/2$, we have a contradiction.

In classical abduction [21], there is a desire to obtain minimalistic explanations— explanations that make the fewest possible assumptions. Defined in Section 2.2.1, the MINSEP problem aheres to this desire. This problem is obviously NP-hard by Theorem 2.1. We can adjust STRAIGHTFORWARD-SEP to find a solution to MINSEP by finding the minimum hitting set of the P_o's.

Example 2.6. Consider the burglar example scenario in Example 2.4 and Figure 2.3. Burglary location (observation) o_1 can be partnered with two possible safe houses $\{p_1, p_2\}$—likewise, burglary location o_2 can be partnered with $\{p_2, p_3\}$. We immediately see that the potential safe house located at p_2 is in both sets. Therefore, p_2 is an explanation for both crime scenes. As this is the only such point, we conclude that $\{p_2\}$ is the minimum-sized solution for the **SEP** problem and hence the only solution of MINSEP.
While it is theoretically possible for STRAIGHTFORWARD-SEP to return this set, there are no assurances it does. As we saw in Example 2.4, $\mathcal{E} = \{p_1, p_2\}$ is a solution to **SEP**, although a solution with lower cardinality ($\{p_2\}$) exists. This is why we introduce the MINSEP problem.

With the complexity of k-SEP, the following corollary tells us the complexity class of the Cost-based Explanation problem. We show this reduction by simply setting the cost function $\chi(\mathcal{E}) = |\mathcal{E}|$.

Corollary 2.1. *Cost-based Explanation is NP-Complete.*

Proof. CLAIM 1: Cost-based Explanation is in the complexity class NP.
This follows directly from Theorem 2.1, instead of checking the size of $\mathcal{E}$, we only need to apply the function χ to the $\mathcal{E}$ produced by the non-deterministic algorithm to ensure that $\chi(\mathcal{E}) \leq v$.

CLAIM 2: Cost-based Explanation is NP-Hard.
We show k-SEP$\leq_p$ CBE. Given an instance of k-SEP, we transform it into an instance of CBE in polynomial time where $\chi(\mathcal{E}) = |\mathcal{E}|$ and $v = k$.

CLAIM 2.1: If there is a set $\mathscr{E}$ such that $\chi(\mathscr{E}) \leq v$ then $|\mathscr{E}| \leq k$.
Straightforward.

CLAIM 2.2: If there is a set $\mathscr{E}$ of size k or less then $\chi(\mathscr{E}) \leq v$
Straightforward.

MINSEP has the feel of a set-covering problem. Although the generalized cost-based explanation cannot be directly viewed with a similar intuition (as the cost maps explanations to reals—not elements of $\mathscr{S}$), there is an important variant of the cost-based problem that does. We introduce weighted **SEP**, or **WT-SEP** below.

Weighted Spatial Explanation. (WT-SEP)
INPUT: A space $\mathscr{S}$, a set $\mathscr{O}$ of observations, a feasibility predicate feas, numbers $0 \leq \alpha < \beta \leq 1$, a weight function $c : \mathscr{S} \to \mathfrak{R}$, and a real number $v > 0$.
OUTPUT: "Yes" if there exists an (α, β) explanation $\mathscr{E}$ for $\mathscr{O}$ such that $\sum_{p \in \mathscr{E}} c(p) \leq v$, and "no" otherwise.

In this case, we can easily show NP-Completeness by reduction from k-**SEP**, we simply set the weight for each element of $\mathscr{S}$ to be one, causing $\sum_{p \in \mathscr{E}} c(p)$ to equal the cardinality of $\mathscr{E}$.

Corollary 2.2. *WT-SEP is NP-Complete.*

Proof. Membership in the complexity class NP follows directly from Theorem 2.1, instead of checking the size of $\mathscr{E}$, we check if $\sum_{p \in \mathscr{E}} c(p) \leq v$. We also note that the construction for cost-based explanation in Theorem 2.1 is also an instance of WT-SEP, hence NP-hardness follows immediately.

Cost-based explanation problems presented in this section are very general because the cost function is not "nailed down". A researcher or reader might wonder if things become any easier if the cost function is a specific cost function rather than being a completely general, arbitrary cost function. We now specify a couple of cost functions below and the associated version of **SEP**.

The total-distance minimization explanation problem (**TD-SEP**) says, intuitively, that given an explanation $\mathscr{E}$ for a given set $\mathscr{O}$ of observations, we should consider the distance between each observation o and its nearest neighbor o_e. The cost of explanation $\mathscr{E}$ is the sum of these distances. We want to find the explanation $\mathscr{E}$ that minimizes this cost, while still insisting that $\mathscr{E}$ has k elements or less.

Total Distance Minimization Explanation Problem. (TD-SEP)
For space $\mathscr{S}$, let $d : \mathscr{S} \times \mathscr{S} \to \mathfrak{R}$ be the Euclidean distance between two points in $\mathscr{S}$.
INPUT: A space $\mathscr{S}$, a set $\mathscr{O}$ of observations, a feasibility predicate feas, numbers $0 \leq \alpha < \beta \leq 1$, positive integer $k < |\mathscr{O}|$, and real number $v > 0$.
OUTPUT: "Yes" if there exists an (α, β) explanation $\mathscr{E}$ for $\mathscr{O}$ such that $|\mathscr{E}| \leq k$ and $\sum_{o_i \in \mathscr{O}} \min_{p_j \in \mathscr{E}} d(o_i, p_j) \leq v$, and "no" otherwise.

Theorem 2.2. *TD-SEP is NP-Complete.*

Proof. CLAIM 1: TD-SEP is in the complexity class NP.

Given a set $\mathcal{E}$, we can easily determine in polynomial time that it meets the standards of the output specified in the problem statement.

CLAIM 2: TD-SEP is NP-hard.

Consider Euclidean k-Median Problem, as presented and shown to be NP-Complete in [19], defined as follows:

INPUT: A set P of integer-coordinate points in a Euclidean plane, positive integer $k' < |P|$, real number $v' > 0$.

OUTPUT: "Yes" if there is a set of points, $S \subseteq P$ such that $|S| = k'$ and $\sum_{x_i \in X} \min_{s_j \in S} d(x_i, s_j) \leq v'$—"no" otherwise.

Given an instance of the Euclidean k-Median Problem, we create an instance of TD-SEP as follows:

- Set $\mathcal{S}$ to be a set of lattice points in the Euclidean plane that include all points in P
- Set $\mathcal{O} = P$
- Let $feas(x) = \text{TRUE}$ iff $x \in P$
- Set $\alpha = 0$
- Set β greater than the diagonal of $\mathcal{S}'$
- Set $k = k'$
- Set $v = v'$

CLAIM 2.1: If there is $\mathcal{E}$, a k-sized explanation for $\mathcal{O}$ such that $\sum_{o_i \in \mathcal{O}} \min_{p_j \in \mathcal{E}} d(o_i, p_j) \leq v$, then there is a set $S \subseteq P$ such that $|S| = k'$ and $\sum_{x_i \in P} \min_{s_j \in S} d(x_i, s_j) \leq v'$.

Because of how we set feas and $\mathcal{O}$, $\mathcal{E} \subseteq P$. As α and β do not affect $\mathcal{E}$, the only real restrictions on $\mathcal{E}$ is that its cardinality is k and that $\sum_{o_i \in \mathcal{O}} \min_{p_j \in \mathcal{E}} d(o_i, p_j) \leq v$. Because of how we set k and v, we can see that $\mathcal{E}$ meets all the conditions to be a solution to the Euclidean k-Median problem, hence the claim follows.

CLAIM 2.2: If there is set $S \subseteq P$ such that $|S| = k'$ and $\sum_{x_i \in P} \min_{s_j \in S} d(x_i, s_j) \leq v'$, then there is set $\mathcal{E}$, a k-sized explanation for $\mathcal{O}$ such that $\sum_{o_i \in \mathcal{O}} \min_{p_j \in \mathcal{E}} d(o_i, p_j) \leq v$.

In the construction, the arguments α, β and feas allow any element of a solution to the k-Median problem to be a partner for any observation in $\mathcal{O}$. By how we set k and v, we can easily see that S is a valid solution to TD-SEP. The claim follows.

The statement of the theorem follows directly from claims 1-2.

The NP-hardness of **TD-SEP** is based on a reduction from the k-Median Problem [19]. The k-median classification algorithm, based on the k-median problem, is a popular way of clustering data. Unfortunately, the k-median problem makes no provision for the notion of "feasibility". As we can see from the above reduction, it is clear that the k-median problem is a special case of GAPs, but k-median problems cannot handle arbitrary feasibility predicates of the kind that occur in real-life geospatial reasoning. The same argument applies to k-means classifiers as well.

2.3.1 Counting Solutions to a GAP

In Theorem 2.1, we showed that the problem of finding an explanation of size k to be NP-Complete based on a reduction from the known NP-Complete problem *Geometric Covering by Discs* (GCD) seen in [14]. As with most decision problems, we define the associated counting problem, #GCD, as the number of "yes" answers to the GCD decision problem. The result below shows that #GCD is #P-complete and, moreover, that there is no fully-polynomial random approximation scheme (or FPRAS for short) for #GCD unless *NP* equals the complexity class *RP*.[3]

Lemma 2.1. *#GCD is #P-complete and has no FPRAS unless NP=RP.*

Proof. **CLAIM 1:** #GCD is in #P.
Clearly, as the total number of "yes" answers is bounded by 2^K, this problem is in the complexity class #P.
CLAIM 2: #GCD is #P-hard.
We have to show a parsimonious or weakly parsimonious reduction from a known #P-complete problem. In [4], the authors show that the counting version of the dominating set problem (#DOMSET) is #P-complete based on a weakly parsimonious reduction from the counting version of vertex cover. It is important to note that the construction used in this proof uses a graph with a maximum degree of three. This shows that the counting version of the dominating set problem on a graph with a maximum degree of three is also #P-hard as well. In [5], the authors show a parsimonious reduction from the dominating set problem (with maximum degree of three) to GCD. Hence, as the reduction is parsimonious, and the associated counting problem is #P-hard, then the statement of the claim follows.
CLAIM 3: There is no FPRAS for #GCD unless NP = RP.
By Lemma 2.1 and [4], consider the following chain of polynomial-time parsimonious or weakly parsimonious reductions:

$$\#SAT \rightarrow \#3CNFSAT \rightarrow \#Pl3CNFSAT$$
$$\#Pl3CNFSAT \rightarrow \#Pl1Ex3SAT \rightarrow \#Pl1Ex3MonoSAT$$
$$\#Pl1Ex3MonoSAT \rightarrow \#PlVC \rightarrow \#Pl3DS \rightarrow \#GCD$$

---

[3] *RP* is the class of decision problems for which there is a randomized polynomial algorithm that, for any instance of the problem, returns "false" with probability 1 when the correct answer to the problem instance is false, and returns "true" with probability $(1 - \varepsilon)$ for a small $\varepsilon > 0$ when the correct answer to the problem instance is "true."

Hence, as all of the reductions are PTIME, parsimonious or weakly parsimonious, and planarity preserving (for planar problems), by the results of [6], the statement follows.

We can leverage the above result to derive a complexity result for the counting version of k-**SEP**.

Theorem 2.3. *The counting version of k-SEP is #P-Complete and has no FPRAS unless NP=RP.*

2.4 Exact Algorithm for GAP Problems

This section presents four exact approaches to solve k-**SEP** and **WT-SEP**. First, we provide an enumerative approach that exhaustively searches for an explanation. We then show that the problem reduces to some well known combinatorial optimization problems. In particular, it reduces to the set-cover, dominating set, and linear-integer programming problems. Once k-**SEP** and **WT-SEP** are reduced to these problems, we can use existing algorithms for these problems to solve k-**SEP** and **WT-SEP**. In this section, we will show these possible ways of solving k-**SEP** and **WT-SEP**.

Throughout this section, we use Δ to represent the bound on the number of partners that can be associated with a single observation and f to represent the bound on the number of observations supported by a single partner. Note that both values are bounded by $\pi(\beta^2 - \alpha^2)$, however they can be much less in practice—specifically f is normally much smaller than Δ.

2.4.1 Naive Exact Algorithm

We now show correctness of NAIVE-KSEP-EXACT. This algorithm provides an exact solution to k-**SEP** but takes exponential time (in k). The algorithm first identifies a set L of all elements of $\mathscr{S}$ that could be possible partners for $\mathscr{O}$. Then, it considers all subsets of L of size less than or equal to k. It does this until it identifies one such subset as an explanation.

Proposition 2.1. *If there is a k-sized simple (α, β) explanation for $\mathscr{O}$, then NAIVE-KSEP-EXACT returns an explanation. Otherwise, it returns NO.*

Proof. CLAIM 1: If there is a k-sized simple (α, β) explanation for $\mathscr{O}$, then NAIVE-KSEP-EXACT returns an explanation.
Suppose, by way of contradiction, that there is a k-sized simple (α, β) explanation for $\mathscr{O}$ and NAIVE-KSEP-EXACT returns NO. Then there does not exist k bit strings such that for all o_i, $\sum_{j=1}^{k}(\ell_j(i)) \geq 1$. As each bit string is associated with a point in $\mathscr{S}$, then by the construction of the bit strings, there are not k points in $\mathscr{S}$ such that

Algorithm 3 (NAIVE-KSEP-EXACT)

INPUT: Space $\mathscr{S}$, a set $\mathscr{O}$ of observations, a feasibility predicate feas, real numbers $\alpha \geq 0$, $\beta > 0$, and natural number $k > 0$

OUTPUT: Set $\mathscr{E} \subseteq \mathscr{S}$ of size k (or less) that explains $\mathscr{O}$

1. Let M be a matrix array of pointers to binary string $\{0,1\}^{|\mathscr{O}|}$. M is of the same dimensions as $\mathscr{S}$. Each element in M is initialized to NULL. For a given $p \in \mathscr{S}$, $M[p]$ is the place in the array.
2. Let L be a list of pointers to binary strings. L is initialized as null.
3. For each $o_i \in \mathscr{O}$ do the following

    a. Determine all points $p \in \mathscr{S}$ such that $\alpha \leq d(o,p) \leq \beta$ such that feas(p) = TRUE.
    b. For each of these points, p, if $M[p]$ = NULL then initialize a new array where only bit i is set to 1. Then add a pointer to $M[p]$ in L.
    c. Otherwise, set bit i of the existing array to 1.

4. For any k elements of L (actually the k elements pointed to by elements of L), we shall designate $\ell_1, \ldots, \ell_j, \ldots \ell_k$ as the elements. We will refer to the ith bit of element ℓ_j as $\ell_j(i)$.
5. Exhaustively generate all possible combinations of k elements of L until one such combination is found where $\forall i \in [1, |\mathscr{O}|], \sum_{j=1}^{k}(\ell_j(i)) > 0$
6. If no such combination is found, return NO. Otherwise, return the first combination that was found.

each point is feasible and falls no closer than α and no further than β distance away from each point in $\mathscr{O}$. This is a contradiction.

CLAIM 2: If there is no k-sized simple (α, β) explanation for $\mathscr{O}$, then NAIVE-KSEP-EXACT returns NO.

Suppose, by way of contradiction, that there is no k-sized simple (α, β) explanation for $\mathscr{O}$ and NAIVE-KSEP-EXACT returns an explanation. Then there must exist k bit strings such that $\bigvee_{j=1}^{k}(\ell_j(i)) = 1$. As each bit string is associated with a point in $\mathscr{S}$, then by the construction of the bit strings, there must exist k points in $\mathscr{S}$ such that each point is feasible and falls no closer than α and no further than β distance away from each point in $\mathscr{O}$. This is a contradiction.

We now state the complexity of the NAIVE-KSEP-EXACT algorithm in the proposition below.

Proposition 2.2. *The complexity of NAIVE-KSEP-EXACT is* $O(\frac{1}{(k-1)!}(\pi(\beta^2 - \alpha^2)|\mathscr{O}|)^{(k+1)})$.

Proof. Note that as all pointers in M are initially null, thus there is no need to iterate through every element in M - rather lists in M can only be initialized as needed. Hence, the cost to set-up M in $O(1)$ and not the size of the matrix.

As each $o \in \mathscr{O}$ has, at most $\pi(\beta^2 - \alpha^2)$ partners, the total complexity of the inner loop is $\pi(\beta^2 - \alpha^2)|\mathscr{O}|$.

As we have, at most, $\pi(\beta^2 - \alpha^2)|\mathscr{O}|$ elements in L (recall that L is the subset of $\mathscr{S}$ that can be partnered with elements in $\mathscr{O}$), then there are $\binom{\pi(\beta^2-\alpha^2)|\mathscr{O}|}{k}$ iterations taking place in step 5. Each iteration costs $k \cdot |\mathscr{O}|$ as we must compare the $|\mathscr{O}|$ bits of each k bit string. So,

$$\binom{\pi(\beta^2 - \alpha^2)|\mathscr{O}|}{k} \cdot k \cdot |\mathscr{O}|$$

$$= \frac{(\pi(\beta^2 - \alpha^2)|\mathcal{O}|) \cdot (\pi(\beta^2 - \alpha^2)|\mathcal{O}| - 1) \cdot \ldots \cdot (\pi(\beta^2 - \alpha^2)|\mathcal{O}| - (k-1))}{k!} \cdot k \cdot |\mathcal{O}|$$

$$< O(\frac{1}{(k-1)!}(\pi(\beta^2 - \alpha^2)|\mathcal{O}|)^{(k+1)})$$

As this term dominates the complexity of the inner loop, the statement follows.

An exact algorithm for the cost-based explanation problems follows trivially from the NAIVE-KSEP-EXACT algorithm by adding the step of computing the value for χ for each combination. Provided this computation takes constant time, this does not affect the $O(\frac{1}{(k-1)!}(\pi(\beta^2 - \alpha^2)|\mathcal{O}|)^{(k+1)})$ run time of that algorithm.

2.4.2 An Exact Set-Cover Based Approach

We now show that k-SEP polynomially reduces to an instance of the popular set-covering problem [15] which allows us to directly apply the well-known greedy algorithm reviewed in [20]. SET_COVER is defined as follows.

The Set-Cover Problem. (SET_COVER)
INPUT: Set of elements E, and a family of subsets of E, $F = \{S_1, \ldots, S_{max}\}$, and positive integer k.
OUTPUT: "Yes" if there exists a k-sized (or less) subset of F, F_k, such that $\bigcup_{S_i \in F_k} S_i = E$.

Through a simple modification of NAIVE-KSEP-EXACT, we can take an instance of k-SEP and produce an instance of SET_COVER. We run the first four steps, which only takes $O(\Delta \cdot |\mathcal{O}|)$ time by the proof of Proposition 2.2. (Recall that Δ represents a bound on the number of partners that can be associated with a single observation).

Theorem 2.4. *k-SEP polynomially reduces to SET_COVER.*

Proof. We employ the first four steps of NAIVE-KSEP-EXACT. We view the bit-strings in list L as subsets of $\mathcal{O}$ where if the ith bit of the string is 1, o_i of $\mathcal{O}$ is in the set.

CLAIM 1: If there are k subsets of L that cover $\mathcal{O}$, then there is a k-sized simple (α, β) explanation for $\mathcal{O}$.
Suppose, by way of contradiction, that there are k subsets of L that cover $\mathcal{O}$ and there is no k-sized simple (α, β) explanation for $\mathcal{O}$. Then, by Proposition 2.1, for every combination of k bit strings, there is some bit i such that $\bigvee_{j=1}^{k}(\ell_j(i)) = 1$ does not hold. Hence, by the reduction, a set cover with k sets from L would be impossible. This is a contradiction.

CLAIM 2: If there there is a k-sized simple (α, β) explanation for $\mathcal{O}$, then there are k subsets of L that cover $\mathcal{O}$.

Suppose, by way of contradiction, there is a k-sized simple (α, β) explanation for $\mathcal{O}$ and there are not k subsets of L that cover $\mathcal{O}$. Then, for any combination of k subsets of L, there is at least one element of $\mathcal{O}$ not included. Hence, for any bit-string representation of an element in L, for some bit i, $\bigvee_{j=1}^{k}(\ell_j(i)) = 1$ does not hold. However, by Proposition 2.1, this must hold or there is no k-sized simple (α, β) explanation for $\mathcal{O}$. This is a contradiction.

We now provide an example of how the NAIVE-KSEP-EXACT algorithm solves the burglar problem.

Example 2.7. Consider the burglar scenario in Example 2.4 and Figure 2.3 (below). Suppose we want to solve this problem as an instance of k-**SEP** by a reduction to set-cover. We consider the set of burglary locations, $\mathcal{O} = \{o_1, o_2\}$ as the set we wish to cover. We obtain our covers from the first four steps of NAIVE-KSEP-EXACT. Let us call the result list L. Hence, we can view the values of the elements in L as the following sets:

1. $S_1 = \{o_1\}$,
2. $S_2 = \{o_1, o_2\}$,
3. $S_3 = \{o_2\}$.

These correspond to the points p_1, p_2, p_3 respectively. As S_2 covers $\mathcal{O}$, p_2 is an explanation.

The traditional approach for approximation of set-cover has a time complexity of $O(|E| \cdot |F| \cdot size)$, where $size$ is the cardinality of the largest set in the family F (*i.e.*, $size = \max_{i \leq |F|} |S_i|$). This approach obtains an approximation ratio of $1 + \ln(size)$ [20]. As f is the quantity of the largest number of observations supported by a single partner, the approximation ratio for k-**SEP** using a greedy-scheme after a reduction from set-cover is $1 + \ln(f)$. The NAIVE-KSEP-SC algorithm below leverages the above reduction to solve the k-SEP problem.

The result belows specifies the complexity of NAIVE-KSEP-SC.

Proposition 2.3. *NAIVE-KSEP-SC has a complexity of $O(\Delta \cdot f \cdot |\mathcal{O}|^2)$ and an approximation ratio of $1 + \ln(f)$. (Recall that Δ is a bound on the number of partners that can be associated with a single observation and f is a bound on the number of observations supported by a single partner).*

Proof. CLAIM 1: NAIVE-KSEP-SC has a complexity of $O(\Delta \cdot f \cdot |\mathcal{O}|^2)$.

The loop at line 3, which reduces the problem to set-covering, takes $O(\Delta \cdot |\mathcal{O}|)$ time.

The loop at line 4 iterates, at most, $|\mathcal{O}|$ times.

The first nested loop at line 4b iterates, at most, $\Delta \cdot |\mathcal{O}|$ times.

The second nested loop at line 4(b)ii iterates, at most, f times.

The updating procedure at line 4d, which is still inside the loop at line 4, iterates, at most, f times.

Algorithm 4 (NAIVE-KSEP-SC)

INPUT: Space $\mathscr{S}$, a set $\mathscr{O}$ of observations, a feasibility predicate feas, and real numbers $\alpha \geq 0, \beta > 0$
OUTPUT: Set $\mathscr{E} \subseteq \mathscr{S}$ that explains $\mathscr{O}$

1. Initialize list $\mathscr{E}$ to null. Let M be a matrix array of the same dimensions as $\mathscr{S}$ of lists of pointers initialized to null. For a given $p \in \mathscr{S}$, $M[p]$ is the place in the array. Let L be a list of pointers to lists in M, L is initialized to null.
2. Let $\mathscr{O}'$ be an array of Booleans of length $|\mathscr{O}|$. $\forall i \in [1, |\mathscr{O}|]$, initialize $\mathscr{O}'[i] = $ TRUE. For some element $o \in \mathscr{O}$, $\mathscr{O}'[o]$ is the corresponding space in the array. Let numObs $= |\mathscr{O}|$
3. For each element $o \in \mathscr{O}$, do the following.

   a. Determine all elements $p \in \mathscr{S}$ such that feas$(p) = $ TRUE and $d(o, p) \in [\alpha, \beta]$
   b. If there does not exist a $p \in \mathscr{S}$ meeting the above criteria, then terminate the program and return IMPOSSIBLE.
   c. If $M[p] = $ null then add a pointer to $M[p]$ to L
   d. Add a pointer to o to the list $M[p]$.

4. While numObs > 0 loop

   a. Initialize pointer cur_ptr to null, integer cur_size to 0
   b. For each $ptr \in L$, do the following:
      i. Initialize integer $this\_size$ to 0, let $M[p]$ be the element of M pointed to by ptr
      ii. For each obs_ptr in the list $M[p]$, do the following
         A. Let i be the corresponding location in array $\mathscr{O}'$ to obs_ptr
         B. If $\mathscr{O}'[i] = $ TRUE, increment $this\_size$ by 1
      iii. If $this\_size > cur\_size$, set $cur\_size = this\_size$ and have cur_ptr point to $M[p]$
   c. Add p to $\mathscr{E}$
   d. For every obs_ptr in the list pointed to by cur_ptr, do the following:
      i. Let i be the corresponding location in array $\mathscr{O}'$ to obs_ptr
      ii. If $\mathscr{O}'[i]$, then set it to FALSE and decrement numObs by 1
   e. Add the location in space $\mathscr{S}$ pointed to by cur_ptr to $\mathscr{E}$

5. Return $\mathscr{E}$

Hence, by the above statements, the total complexity of **NAIVE-KSEP-SC** is $O(|\mathscr{O}| \cdot (\Delta \cdot |\mathscr{O}| \cdot f + f) + \Delta \cdot |\mathscr{O}|)$, hence the statement follows.

CLAIM 2: **NAIVE-KSEP-SC** has an approximation ratio of $1 + ln(f)$.

Viewing list L as a family of subsets, each subset is the set of observations associated with a potential partner, hence the size of the subsets is bounded by f. The approximation ratio follows directly from the analysis of the set-covering problem.

The result below shows that **NAIVE-KSEP-SC** solves the k-SEP problem.

Proposition 2.4. *A solution $\mathscr{E}$ to* **NAIVE-KSEP-SC** *provides a partner to every observation in $\mathscr{O}$ if a partner exists–otherwise, it returns IMPOSSIBLE.*

Proof. Follows directly from Theorem 2.4.

The algorithm **NAIVE-KSEP-SC** is a naive, straightforward application of the $O(|E| \cdot |F| \cdot size)$ greedy approach for set-cover as presented in [20]. We note that it is possible to implement a heap to reduce the time-complexity to $O(\Delta \cdot f \cdot |\mathscr{O}| \cdot \lg(\Delta \cdot |\mathscr{O}|))$—avoiding the cost of iterating through all possible partners in the inner-loop.

In addition to the straightforward greedy algorithm for set-covering, there are several other algorithms that provide different time complexity/approximation ratio combinations. However, with a reduction to the set-covering problem we must consider the result of [18] which states that set-cover cannot be approximated within a ratio $c \cdot log(n)$ for any $c < 0.25$ (where n is the number of subsets in the family F) unless $NP \subseteq DTIME[n^{\text{poly log } n}]$.

Algorithm 5 (KSEP-TO-DOMSET)

INPUT: Space $\mathscr{S}$, a set $\mathscr{O}$ of observations, a feasibility predicate feas, and real numbers $\alpha \geq 0$, $\beta > 0$

OUTPUT: Graph $G_{\mathscr{O}}$ for use in an instance of a **DomSet** problem

1. Let $G_{\mathscr{O}} = (V_{\mathscr{O}}, E_{\mathscr{O}})$ be a graph. Set $V_{\mathscr{O}} = \mathscr{S}$ and $E_{\mathscr{O}} = \emptyset$.
2. Let S be a mapping defined as $S : \mathscr{S} \to V_{\mathscr{O}}$. In words, S takes elements of the space and returns nodes from $G_{\mathscr{O}}$ as defined in the first step. This mapping does not change during the course of the algorithm.
3. For each $o_i \in \mathscr{O}$ do the following

   a. Determine all points $p \in S$ that are such that $\alpha \leq d(o, p) \leq \beta$. Call this set P_i
   b. For all $p \in P_i$ calculate feas(p). If feas$(p) =$ FALSE, remove p from P_i.
   c. Let $V_i = \{v \in V_{\mathscr{O}} | \exists p \in P_i \text{ such that } S(p) = v\}$.
   d. Add $|P_i|$ *new* nodes to $V_{\mathscr{O}}$. Add these nodes to V_i as well.
   e. For every pair of nodes $v_1, v_2 \in V_i$, add edge (v_1, v_2) to $E_{\mathscr{O}}$.

4. Remove all $v \in V_{\mathscr{O}}$ where there does not exist an v' such that $(v, v') \in E_{\mathscr{O}}$
5. If any $P_i = \emptyset$ return IMPOSSIBLE. Otherwise return $G_{\mathscr{O}}$.

A reduction to set-covering has the advantage of being straightforward. It also allows us to leverage the wealth of approaches developed for this well-known problem. In the next section, we show that k-**SEP** reduces to the dominating set problem as well. We then explore alternate approximation techniques based on this reduction.

2.4.3 An Exact Dominating Set Based Approach

We show below that k-**SEP** also reduces to the well known dominating set problem (**DomSet**) [9] allowing us to potentially leverage fast algorithms such as the randomized-distributed approximation scheme in [13]. **DomSet** is defined as follows.

Dominating Set. (DomSet)
INPUT: Graph $G = (V, E)$ and positive integer $K \leq |V|$.
OUTPUT: "Yes" if there is a subset $V' \subset V$ such that $|V'| \leq K$ and such that every vertex $v \in V - V'$ is joined to at least one member of V' by an edge in E.

As the dominating set problem relies on finding a certain set of nodes in a graph, then, unsurprisingly, our reduction algorithm, Algorithm 5, takes space $\mathscr{S}$, an observation set $\mathscr{O}$, feasibility predicate feas, and numbers α, β and returns graph $G_{\mathscr{O}}$ based on these arguments.

We now present an example to illustrate the relationship between a dominating set of size k in $G_{\mathscr{O}}$ and a k-sized simple (α, β) explanation for $\mathscr{O}$. The following example illustrates the relationship between a k-**SEP** problem and **DomSet**.

Example 2.8. Consider the burglar scenario of Example 2.4 as shown in Figure 2.3 (below). Suppose we want to solve this problem as an instance of k-**SEP** by a reduction to **DomSet**. We want to find a 1-sized simple (α, β) explanation (safe-house) for $\mathcal{O}$ (the set of locations where the burglaries occurred, $\{o_1, o_2\}$). Suppose that after running an algorithm such as STRAIGHTFORWARD-SEP, we find that $\{p_1, p_2, p_3\}$ are elements of $\mathcal{S}$ that are feasible. $\{p_1, p_2\}$ are all within a distance of α, β from o_1 and $\{p_2, p_3\}$ are all within a distance of α, β from o_2. We run KSEP-TO-DOMSET which creates graph, $G_{\mathcal{O}}$. This graph is shown in Figure 2.5. We can see that $\{p_2\}$ is a 1-sized dominating sets for $G_{\mathcal{O}}$; hence it is a 1-sized explanation for $\mathcal{O}$.

Fig. 2.5 Results of KSEP-TO-DOMSET based on data seen in Figure 2.3. Note that $\{p_1, p_2, p_1', p_2'\}$ form a complete graph and $\{p_2, p_3, p_2'', p_3'\}$ also form a complete graph. Moreover, $\{p_2\}$ is a dominating set of size 1. Hence, $\{p_2\}$ is a 1-sized simple (α, β) explanation for $\mathcal{O}$, as depicted in Figure 2.3 (below).

We note that the inner loop of KSEP-TO-DOMSET is bounded by $O(\Delta)$ operations and the outer loop will iterate $|\mathcal{O}|$ times. Thus, the complexity of KSEP-TO-DOMSET is $O(\Delta \cdot |\mathcal{O}|)$ as stated in the result below.

Proposition 2.5. *The complexity of KSEP-TO-DOMSET is $O(\Delta \cdot |\mathcal{O}|)$.*

Proof. Notice that the number of points in $\mathcal{S}$ considered for each $o \in \mathcal{O}$ examined in the inner loop is bounded by $O(\Delta)$. As the outer loop is bounded by the size of $\mathcal{O}$, the complexity of KSEP-TO-DOMSET is $O(|\mathcal{O}|)$.

Example 2.8 should give us some intuition into why the reduction to **DomSet** works. The following result states that the k-**SEP** problem can be reduced to **Dom-Set** in polynomial time.

Theorem 2.5. *k-SEP is polynomially reducible to DomSet.*

Proof. We can run KSEP-TO-DOMSET that creates graph $G_{\mathcal{O}} = (V_{\mathcal{O}}, E_{\mathcal{O}})$ based on the set of observations. We show that $G_{\mathcal{O}}$ has a dominating set of size k if and only if there is a k-sized simple (α, β) explanation for $\mathcal{O}$.

CLAIM 1: If $G_{\mathcal{O}}$ has a dominating set of size k or less, then there is a k-sized (or fewer) simple (α, β) explanation for $\mathcal{O}$.
Suppose, by way of contradiction, that $G_{\mathcal{O}}$ has a dominating set of size k and there

is not a k-sized simple (α, β) explanation for $\mathscr{O}$. Then, there has to be at least one element $o_i \in \mathscr{O}$ such that there is no feasible $p \in \mathscr{S}$ where $\alpha \leq d(o_i, p) \leq \beta$. Consider the nodes V_i from the inner loop of KSEP-TO-DOMSET that are associated with o_i. Note that these nodes form a complete subgraph. As each node in V_i is associated with o_i, no node in V_i can be in the dominating set of $G_{\mathscr{O}}$ (if one were, then we would have a contradiction). However, note that half of the nodes in V_i only have edges to other nodes in V_i, so there must be an element of V_i in the dominating set. This is a contradiction.

CLAIM 2: If there is a k-sized simple (α, β) explanation for $\mathscr{O}$, then $G_{\mathscr{O}}$ has a dominating set of size k or less.

Suppose, by way of contradiction, that there is a k-sized simple (α, β) explanation for $\mathscr{O}$, and $G_{\mathscr{O}}$ has does not have a dominating set of size k or less. Let $\mathscr{E}$ be a k-sized simple (α, β) explanation for $\mathscr{O}$. Let this also be a subset of the nodes in $G_{\mathscr{O}}$. By the KSEP-TO-DOMSET, in each set of nodes V_i, there must be at least one element of $\mathscr{E}$. As each set of vertices V_i is a complete graph, then we have a dominating set of size k. Hence, a contradiction.

The straightforward approximation scheme for **DomSet** is to view the problem as an instance of SET_COVER and apply a classical greedy algorithm for SET_COVER. The reduction would view the set of vertices in $G_{\mathscr{O}}$ as the elements, and the family of sets as each vertex and its neighbors. This results in both a greater complexity and a worse approximation ratio when compared with the reduction directly to SET_COVER. This is shown in the following result.

Proposition 2.6. *Solving k-SEP by a reduction to **DomSet** using a straightforward greedy approach has time-complexity $O(\Delta^3 \cdot f \cdot |\mathscr{O}|^2)$ and an approximation ratio bounded by $O(1 + \ln(2 \cdot f \cdot \Delta))$.*

Proof. This is done by a well known reduction of an instance of **DomSet** into an instance of SET_COVER. In the reduction, each node is an element, and the subsets are formed by each node and its neighbors. The Table 2.1 shows the quantities:

Item	Quantity
Number of elements to be covered (number of nodes in $G_{\mathscr{O}}$)	$2 \cdot \Delta \cdot |\mathscr{O}|$
Number of subsets (number of nodes in $G_{\mathscr{O}}$)	$2 \cdot \Delta \cdot |\mathscr{O}|$
Number of elements per subset (Maximum degree of nodes in $G_{\mathscr{O}}$ determined by the produce of partners per observation and observations per partner	$2 \cdot \Delta \cdot f$

Table 2.1 Quantities for the Greedy-Approach in the **DomSet** reduction.

Hence, the total time complexity of the algorithm is $O(8 \cdot \Delta^3 \cdot f \cdot |\mathscr{O}|^2)$ and the complexity part of the statement follows. As the maximum number of elements per subset is $2 \cdot \Delta \cdot f$, the approximation ratio $O(1 + \ln(2 \cdot f \cdot \Delta))$ follows by the well known analysis of the greedy set-covering algorithm.

There are other algorithms to approximate **DomSet** [13, 16]. By leveraging [13], we can obtain an improved complexity while retaining the same approximation ratio as the greedy approach. This is shown below.

Proposition 2.7. *Solving k-SEP by a reduction to DomSet using the distributed, randomized algorithm presented in [13] has a time complexity $O(\Delta \cdot |\mathcal{O}| + \ln(2 \cdot \Delta \cdot |\mathcal{O}|) \cdot \ln(2 \cdot \Delta \cdot f))$ with high probability and approximation ratio of $O(1 + \ln(2 \cdot f \cdot \Delta))$.*

Proof. By Proposition 2.5, the complexity of KSEP-TO-DOMSET is $O(\Delta \cdot |\mathcal{O}|)$. The graph $G_{\mathcal{O}}$ has $O(2 \cdot \Delta \cdot |\mathcal{O}|)$ nodes, and the maximum degree of each node is bounded $2 \cdot \Delta \cdot f$ as per Proposition 2.6. As the algorithm in [13] has a complexity of $O(lg(n) \cdot lg(d))$ (with high probability) where n is the number of nodes and d is the maximum degree, the complexity of this approach requires $O(\Delta \cdot |\mathcal{O}| + ln(2 \cdot \Delta \cdot |\mathcal{O}|) \cdot ln(2 \cdot \Delta \cdot f))$ with high probability (the statement follows).

As the approach in [13] is greedy, it maintains the $O(1 + \ln(2 \cdot f \cdot \Delta))$ (Proposition 2.6) (the approximation ratio in this case being a factor of the optimal in expectation).

Hence, although a reduction to dominating set generally gives us a worse approximation guarantee, we can (theoretically) outperform set-cover with the randomized algorithm for dominating set in terms of complexity.

2.4.4 An Exact Integer Linear Programming based Approach

Given an instance of k-**SEP**, this section shows how to create a set of integer constraints that, if solved, will yield a solution to the problem. We start by defining the set of integer linear constraints associated with k-**SEP**.

Definition 2.8 (OPT-KSEP-IPC). The k-**SEP** integer programming constraints (OPT-KSEP-IPC) require the following information, obtained in $O(|\mathcal{O}| \cdot \pi(\beta^2 - \alpha^2))$ time:

- Let L be the set of all possible partners generated in the first four steps of NAIVE-KSEP-EXACT.
- For each $p \in L$, let $str(p)$ be the string of $|\mathcal{O}|$ bits, where bit $str(p)_i$ is 1 if p is a partner of the ith observation (this is also generated in the first four steps of NAIVE-KSEP-EXACT).

For each $p_j \in L$, we use the variable $x_j \in \{0, 1\}$ as follows. We would like $x_j = 1$ iff p_j is in $\mathcal{E}$. Then KSEP-IPC consists of the following:
Minimize $\sum_{p_j \in L} x_j$ subject to

1. $\forall o_i \in \mathcal{O}, \sum_{p_j \in L} x_j \cdot str(p_j)_i \geq 1$
2. $\forall p_j \in L, x_j \in \{0,1\}$ (later, we will relax the integer constraint on variables in the integer program and use the constraint $x_j \leq 1$)

The following result describes the size of the above integer linear program.

Proposition 2.8. *OPT-KSEP-IPC consists of* $O(|\mathcal{O}|\pi(\beta^2 - \alpha^2))$ *variables and* $O(|\mathcal{O}| \cdot \pi(\beta^2 - \alpha^2))$ *constraints.*

Proof. Follows directly from Definition 2.8.

The result below shows that the above integer linear program does in fact correctly capture the desired solution(s) to k-**SEP**.

Proposition 2.9. *For a given instance of the optimization version k-SEP, if OPT-KSEP-IPC is solved, then* $\bigcup_{p_j \in L_{x_j = 1}} p_j$ *is an optimal solution to k-SEP.*

Proof. Suppose, by way of contradiction, that $\bigcup_{p_j \in L_{x_j = 1}} p_j$ is not an optimal solution to k-**SEP**. By the constraint, $\forall o_i \in \mathcal{O}, \sum_{p_j \in L} x_j \cdot str(p_j)_i \geq 1$, we are ensured that for each observation, there is a partner p_j such that $x_j = 1$. Further, if we associate x_j with the selected parter p_j for any solution $\mathcal{E}$ to k-**SEP**, then this constraint must hold. Hence, $\bigcup_{p_j \in L_{x_j = 1}} p_j$ is a valid explanation. Therefore, the optimal solution to the instance of k-**SEP**, we shall call $\mathcal{E}_{OPT}$, must be smaller than $\bigcup_{p_j \in L_{x_j = 1}} p_j$. As the minimization of $\sum_{p_j \in L} x_j$ ensures that the cardinality of $\bigcup_{p_j \in L_{x_j = 1}} p_j$ is minimized. Therefore, $|\mathcal{E}_{OPT}|$ cannot be smaller than $|\bigcup_{p_j \in L_{x_j = 1}} p_j|$, as the constraint $\forall o_i \in \mathcal{O}$, $\sum_{p_j \in L} x_j \cdot str(p_j)_i \geq 0$ holds for any solution to k-**SEP**. This is a contradiction.

We now return to our burglar example.

Example 2.9. Consider the burglar scenario in Example 2.4, pictured in Figure 2.3 (below). Suppose we want to solve this problem as an instance of MINSEP. We would set up the constraints as follows:
Minimize $x_1 + x_2 + x_3$ **subject to** $1 \cdot x_1 + 1 \cdot x_2 + 0 \cdot x_3 \geq 1$ and $0 \cdot x_1 + 1 \cdot x_2 + 1 \cdot x_3 \geq 1$, where $x_1, x_2, x_3 \in \{0,1\}$
Obviously, setting $x_1 = 0, x_2 = 1, x_3 = 0$ provides an optimal solution. Hence, as x_2 is the only non-zero variable, p_2 is the explanation for the crime-scenes.

A solution to the constraints OPT-KSEP-IPC can be approximated using the well-known "rounding" technique [10, 32] that relaxes constraints. We present an OPT-KSEP-IPC using rounding.

Proposition 2.10. *NAIVE-KSEP-ROUND returns an explanation for $\mathcal{O}$ that is within a factor Δ from optimal, where Δ is the maximum number of possible partners associated with any observation.*

Algorithm 6 (NAIVE-KSEP-ROUND)

INPUT: Space $\mathscr{S}$, a set $\mathscr{O}$ of observations, a feasibility predicate feas, and real numbers $\alpha \geq 0$, $\beta > 0$

OUTPUT: Set $\mathscr{E} \subseteq \mathscr{S}$ that explains $\mathscr{O}$

1. Run the first four steps of NAIVE-KSEP-EXACT
2. Solve the relaxation of OPT-KSEP-IPC
3. For the $o \in \mathscr{O}$ with the most possible partners, let Δ be the number of possible partners associated with o. (This can be done in line 1).
4. Return all $p_j \in L$ where $x_j \geq \frac{1}{\Delta}$

There are several things to note about this approach. First, it can be easily adapted to many of the weighted variants—such as **WT-SEP**. Second, we note that the rounding algorithm is not a randomized rounding algorithm—which often produces a solution that satisfies all of the constraints in the linear-integer program. The above algorithm guarantees that all of the observations will be covered (if an explanation exists). Finally, this approach allows us to leverage numerous software packages for solving linear and linear-integer programs.

2.5 Greedy Heuristics for GAP Problems

In this section, we suggest some greedy approaches to solve geospatial abduction problems.

2.5.1 A Linear Time Greedy Approximation Scheme

In this section, we introduce a greedy approximation scheme for the optimization version of k-**SEP** that has a lower time-complexity than NAIVE-KSEP-SC but still maintains the same approximation ratio. Our GREEDY-KSEP-OPT1 algorithm runs in linear time with respect to $\mathscr{O}$. The key intuition is that NAIVE-KSEP-SC iterates through $O(\Delta \cdot |\mathscr{O}|)$ possible partners in line 4. Our algorithm first randomly picks an observation and then greedily selects a partner for it. This results in the greedy step iterating through only $O(\Delta)$ partners.

Example 2.10. Consider the tiger kill example from Example 2.3 and Figure 2.3. After initializing the necessary data structures in lines 1-3, GREEDY-KSEP-OPT1 iterates through the observations in $\mathscr{O}$ where the associated position in $\mathscr{O}'$ is TRUE. Suppose the algorithm picks o_1 first. It now accesses the list pointed to from OBS[o_1]. This gives us a set of pointers to the following elements of $\mathscr{S}$: $\{p_1, p_2, p_3, p_4\}$. Following the greedy selection outlined in line 4 of NAIVE-KSEP-

Algorithm 7 (GREEDY-KSEP-OPT1)

INPUT: Space $\mathscr{S}$, a set $\mathcal{O}$ of observations, a feasibility predicate feas, and real numbers $\alpha \geq 0$, $\beta > 0$

OUTPUT: Set $\mathscr{E} \subseteq \mathscr{S}$ that explains $\mathcal{O}$

1. Run lines 1-2 of NAIVE-KSEP-SC
2. Let OBS be an array, size $|\mathcal{O}|$ of lists to pointers in M. For some observation o, let OBS$[o]$ be the corresponding list in the array.
3. Run the loop in line 3 of NAIVE-KSEP-SC but when partner p of observation o is considered, add a pointer to $M[p]$ in the list OBS$[o]$. The list L need not be maintained.
4. While numObs > 0 loop

   a. Randomly select an element $o \in \mathcal{O}$ such that $\mathcal{O}'[o] = $ TRUE
   b. Run the greedy-selection loop of line 4 of NAIVE-KSEP-SC, but consider the list OBS$[o]$ instead of L

5. Return $\mathscr{E}$

SC, the algorithm iterates through these points, visiting the list of observations associated with each one in the matrix array M.

First, the algorithm accesses the list pointed to by $M[p_1]$. Figure 2.6 (above) shows the observations considered when p_1 is selected. As there is only one observation in list $M[p_1]$ whose associated Boolean in $\mathcal{O}'$ is TRUE, the variable *cur_size* is set to 1 (see line 4(b)iii of NAIVE-KSEP-SC). *cur_ptr* is then set to $M[p_1]$.

We now consider the next element, p_2. Figure 2.6 (right) shows the list pointed to by $M[p_2]$. As $M[p_2]$ points to more observations whose associated $\mathcal{O}'$ Boolean is TRUE, we update *cur_size* to 2 and *cur_ptr* to $M[p_2]$.

The algorithm then iterates through p_3 and p_4, but finds they do not offer more observations than p_2. Hence, p_2 is added to the solution set ($\mathscr{E}$). The algorithm updates the array of Booleans, $\mathcal{O}'$ and sets $\mathcal{O}'[o_1]$ and $\mathcal{O}'[o_2]$ to FALSE (depicted by X's over those observations in subsequent figures). numObs is decremented by 2.

We now enter the second iteration of line 4. The only element for the algorithm to pick at this point is o_3, as only $\mathcal{O}'[o_3]$ is TRUE. The list OBS$[o_3]$ points to the positions $\{p_6, p_7, p_8\}$. In Figure 2.7 we look at what happens as the algorithm considers the p_7. As OBS$[o_2] = $ FALSE, it only considers o_3 when computing *this_size*.

When the algorithm finishes its consideration of all the elements pointed to by OBS$[o_3]$, it will return the first element of that set (p_6) as neither p_7 nor p_8 were partners to more available observations than p_6 (in our implementation of this algorithm, we use a coin-flip to break ties among partners with the same number of observations). GREEDY-KSEP-OPT1 then adds p_6 to $\mathscr{E}$ and terminates. The final solution returned, $\{p_2, p_6\}$, is a valid (and in this case, optimal) explanation.

The result below captures the running time of the GREEDY-KSEP-OPT1 algorithm.

Proposition 2.11 (Complexity of GREEDY-KSEP-OPT1). *GREEDY-KSEP-OPT1 has a complexity of $O(\Delta \cdot f \cdot |\mathcal{O}|)$ and an approximation ratio of $1 + \ln(f)$.*

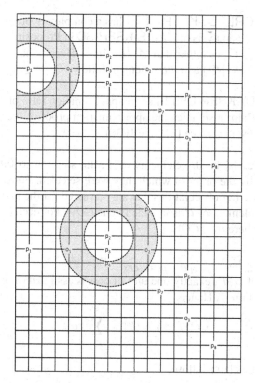

Fig. 2.6 Above: GREEDY-KSEP-OPT1 accesses the list pointed to by $M[p_1]$ thus considering all observations available to p_1. **Below:** GREEDY-KSEP-OPT1 accesses the list pointed to by $M[p_2]$ and finds it has more active observations than it found in the list pointed to by $M[p_1]$.

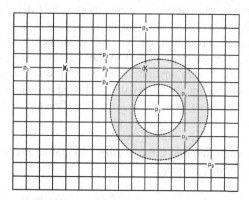

Fig. 2.7 GREEDY-KSEP-OPT1 considers the observations available to p_7. The X's on o_1 and o_2 signify that OBS$[o_1]$ and OBS$[o_2]$ are set to FALSE.

Proof. CLAIM 1: GREEDY-KSEP-OPT1 has a complexity of $O(\Delta \cdot f \cdot |\mathcal{O}|)$.
This follows the same analysis of NAIVE-KSEP-SC in Proposition 2.3, except that
line 4 iterates only Δ times rather than $\Delta \cdot |\mathcal{O}|$ times. Hence, the total complexity is
$O(|\mathcal{O}| \cdot (\Delta \cdot f + f) + \Delta \cdot |\mathcal{O}|)$ and the statement follows.

CLAIM 2: GREEDY-KSEP-OPT1 has an approximation ratio of $1 + ln(f)$.
The proof of this claim resembles the approximation proof of the standard greedy
algorithm for set-cover (*i.e.*, see [3] page 1036).

Let $p_1, \ldots, p_i, \ldots, p_n$ be the elements of $\mathcal{E}$, the solution to GREEDY-KSEP-
OPT1, numbered by the order in which they were selected. For each iteration, let
set COV_i be the subset of observations that are partnered for the first time with point
p_i. Note that each element of $\mathcal{O}$ is in exactly one COV_i. For each $o_j \in \mathcal{O}$, we define
$cost_j$ to be $\frac{1}{|COV_i|}$ where $o_j \in COV_i$.

CLAIM 2.1: $\sum_{p_i \in \mathcal{E}^*} \sum_{o_j \in \mathcal{O}}_{p_i, o_j \text{ are partners}} cost_j \geq |\mathcal{E}|$
By the definition of $cost_j$, exactly one unit of cost is assigned every time a point is
picked for the solution $\mathcal{E}$. Hence,

$$COST(\mathcal{E}) = |\mathcal{E}| = \sum_{o_j \in \mathcal{O}} cost_j$$

The statement of the claim follows.

CLAIM 2.2: For some point $p \in L$, $\sum_{o_j \in \mathcal{O}}_{p, o_j \text{ are partners}} cost_j \leq 1 + ln(f)$.
Let P be the subset of $\mathcal{O}$ that can be partners with p. At each iteration i of the
algorithm, let $uncov_i$ be the number of elements in P that do not have a partner.
Let $last$ be the smallest number such that $uncov_{last} = 0$. Let $\mathcal{E}_P = \{p_i \in \mathcal{E} | (i \leq last) \wedge (COV_i \cap P \neq \emptyset)\}$. From here on, we shall renumber each element in $\mathcal{E}_P$ as
$p_1, \ldots, p_{|\mathcal{E}_P|}$ by the order they are picked in the algorithm (*i.e.*, if an element is
picked that cannot partner with anything in P, we ignore it and continue number-
ing with the next available number, we will use this new numbering for COV_i and
the iterations of the algorithm as well, but do not redefine the set based on the new
numbering).

We note that for each iteration i, the number of items in P that are partnered is
equal to $uncov_{i-1} - uncov_i$. Hence,

$$\sum_{\substack{o_j \in \mathcal{O} \\ p, o_j \text{ are partners}}} cost_j = \sum_{i=1}^{last} \frac{uncov_{i-1} - uncov_i}{|COV_i|}$$

At each iteration of the algorithm, let $PCOV_i$ be the subset of observations that are
covered for the first time if point p is picked instead of point p_i. We note, that
for all iterations in $1, \ldots, last$, the point p is considered by the algorithm as one

of its options for greedy selection. Therefore, as p is not chosen, we know that $|COV_i| \geq |PCOV_i|$. Also, by the definition of $ucov_i$, we know that $|PCOV_i| = ucov_{i-1}$. This gives us:

$$\sum_{\substack{o_j \in \mathcal{O} \\ p, o_j \ are \ partners}} cost_j \leq \sum_{i=1}^{last} \frac{uncov_{i-1} - uncov_i}{ucov_{i-1}}$$

Using the algebraic manipulations of [3] (page 1037), we get the following:

$$\sum_{\substack{o_j \in \mathcal{O} \\ p, o_j \ are \ partners}} cost_j \leq H_{|P|}$$

where H_j is the jth harmonic number. By definition of the symbol f (maximum number of observations supported by a single partner), we obtain the statement of the claim.

(Proof of claim 2): Combining claims 1–2, we get $|\mathcal{E}| \leq \sum_{p_i \in \mathcal{E}^*}(1 + \ln(f))$, which gives us the claim.

The result below establishes the correctness of GREEDY-KSEP-OPT1.

Proposition 2.12. *GREEDY-KSEP-OPT1 returns a $|\mathcal{E}|$-sized (α, β) explanation for $\mathcal{O}$.*
GREEDY-KSEP-OPT1 returns IMPOSSIBLE if there is no explanation for $\mathcal{O}$.

Proof. Suppose by way of contradiction that there exists and element $o \in \mathcal{O}$ such that there is no in $\mathcal{E}$. We note that set $\mathcal{O}'$ contains all elements of $\mathcal{O}$ and the only way for an element to be removed from $\mathcal{O}'$ is if a partner for that element is added to $\mathcal{E}$. Hence, if the program returns a set $\mathcal{E}$, we are guaranteed that each $o \in \mathcal{O}$ has a partner in $\mathcal{E}$.

Suppose by way of contradiction that GREEDY-KSEP-OPT1 returns IMPOS-SIBLE and there exists a set $\mathcal{E}$ that is a valid (α, β) explanation for $\mathcal{O}$. Then, for every element of $\mathcal{O}$, there exists a valid partner. However, this contradicts line 3b of NAIVE-KSEP-SC (called by line 4b of GREEDY-KSEP-OPT1) which causes the program to return IMPOSSIBLE only if an element of $\mathcal{O}$ is found without any possible partner.

We can bound the approximation ratio for GREEDY-KSEP-OPT1 by $O(1 + \ln(f))$, as it is still essentially a greedy algorithm for a covering problem. The main difference between GREEDY-KSEP-OPT1 is the way it greedily chooses covers (partners). This algorithm randomly picks an uncovered observation in each loop and then greedily chooses a cover that covers that observation. Improving the ac-curacy of this algorithm (in practice) is tied directly to the selection criteria used to pick observations, which is random in GREEDY-KSEP-OPT1. In Section 2.5.2

we develop an algorithm that "smartly" picks observations with a dynamic ranking scheme while maintaining a time complexity lower than the standard set-covering approach.

2.5.2 Greedy Observation Selection

GREEDY-KSEP-OPT1 randomly selects observations although subsequent partner selection was greedy. It is easy to implement an *a-priori* ranking of observations based on something like the maximum number of other observations which share a partner with it. Such a ranking could be implemented at the start of GREEDY-KSEP-OPT1 with no effect on complexity, but the ranking would be static and may lose its meaning after several iterations of the algorithm. We could also implement a dynamic ranking. We present a version of GREEDY-KSEP-OPT1 that we call GREEDY-KSEP-OPT2 that picks the observations based on dynamic ranking, runs in time $O(\Delta \cdot f^2 \cdot |\mathscr{O}| + |\mathscr{O}| \cdot \ln(|\mathscr{O}|))$, and maintains the usual approximation ratio of $1 + \ln(f)$ for greedy algorithms. Our key intuition was to use a Fibonacci heap [7]. With such a data structure, we can update the rankings of observations at constant amortized cost per observation being updated. The most expensive operation is to remove an observation from the heap—which costs an amortized $O(\ln(|\mathscr{O}|))$, however as we can never remove more than $|\mathscr{O}|$ items from the heap, this cost is most likely dominated by the cost of the rest of the algorithm, which is more expensive than GREEDY-KSEP-OPT1 by a factor of f. Recall that f is the bound on the number of observations supported by a single partner - and is often very small in practice.

In order to leverage the Fibonacci heap, there are some restrictions on how the ranking can be implemented. First, the heap puts an element with the minimal key on top, and can only decrease the key of elements—an element in the heap can never have its key increased. Additionally, there is a need for some auxiliary data structures as searching for an element in the heap is very expensive. Fortunately, the k-**SEP** problem is amenable to these type of data structures.

We based the key (ranking) on a simple heuristic for each observation. The key for a given observation o is the number of unique observations that share a partner with o. As we are extracting the minimum-keyed observation, we are taking the observation that has the "least in common" with the other observations. The intuition of choosing an observation with "less in common" with other observations ensures that outliers get covered with larger covers. Meanwhile, elements with a higher rank in this scheme are covered last, which may lead to a more efficient cover. In Section 2.6 we show experimentally that this heuristic was viable for the data-set we considered—providing more accurate results than the reduction from set-covering.

Example 2.11. The basic intuition behind GREEDY-KSEP-OPT2 is similar to GREEDY-KSEP-OPT1 in that it iterates through the observations and greedily chooses a partner. The main difference is that it ranks the observations instead of just randomly selecting them. Consider the tiger from Example 2.3 whose behavior is

Algorithm 8 GREEDY-KSEP-OPT2

INPUT: Space $\mathscr{S}$, a set $\mathscr{O}$ of observations, a feasibility predicate feas, and real numbers $\alpha \geq 0$, $\beta > 0$
OUTPUT: Set $\mathscr{E} \subseteq \mathscr{S}$ that explains $\mathscr{O}$

1. Run lines 1-3 of GREEDY-KSEP-OPT1.
2. Let $key_1, \ldots key_{|\mathscr{O}|}$ be natural numbers associated with each observation. Initially, they are set to 0. For some $o \in \mathscr{O}$ let key_o be the associated number.
3. Let REL_OBS be an array of lists of pointers to elements of $\mathscr{O}$. The size of the array is $\mathscr{O}$. For element $o \in \mathscr{O}$, let REL_OBS[o] be the corresponding space in the array.
4. For each $o \in \mathscr{O}$, do the following:

   a. For each element $p \in$ OBS[o], do the following.
      i. For each element obs_ptr of the list pointed to by $M[p]$, do the following
         A. If obs_ptr points to an element of $\mathscr{O}$ not pointed to in the list REL_OBS[o], then add obs_ptr to REL_OBS[o] and increment key_o by 1.

5. Let OBS_HEAP be a Fibonacci heap. Let QUICK_LOOK be an array (size $\mathscr{O}$) of pointers to elements of the heap. For each $o \in \mathscr{O}$, add the tuple $\langle o, key_o \rangle$ to the heap, along with a pointer to the tuple to QUICK_LOOK[o]. Note we are using key_o as the key for each element in the heap.
6. While OBS_HEAP is not empty, loop

   a. Take the minimum element of OBS_HEAP, let o be the associated observation with this element.
   b. Greedily select an element of OBS[o] as done in the loop at line 4 of GREEDY-KSEP-OPT1. We shall call this element p.
   c. For every $o' \in \mathscr{O}$ pointed to by a pointer in $M[p]$, such that $\mathscr{O}'[o'] =$ TRUE, do the following.
      i. Set $\mathscr{O}'[o'] =$ FALSE
      ii. Remove the element pointed to by QUICK_LOOK[o'] from OBS_HEAP
      iii. For every element $o'' \in \mathscr{O}$ pointed to by an element of REL_OBS[o'] where $\mathscr{O}'[o''] =$ TRUE do the following.
         A. Decrease the $key_{o''}$ by 1.

7. Return $\mathscr{E}$

Observation	key_i	REL_OBS[o_i]
o_1	2	$\{o_1, o_2\}$
o_2	2	$\{o_1, o_2\}$
o_3	2	$\{o_2, o_3\}$

Table 2.2 key values and related observations for observations in the tiger kill scenario introduced in Example 2.3.

depicted in Figure 2.3. In Example 2.10, we used GREEDY-KSEP-OPT1 to solve the associated k-**SEP** problem for this situation. We shall discuss how GREEDY-KSEP-OPT2 differs.

The first main difference is that the algorithm assigns a rank to each observation o_i, called key_i, which is also the key used in the Fibonacci heap. This is done in the loop at line 4. It not only calculates key_i for each observation, but it also records the elements "related" to it in the array REL_OBS. Note that a "related" observation needs only to share a partner with a given observation. Not all related observations need to have the same partner. For the tiger kill scenario, we show the keys and related observations in Table 2.2.

As the key values are the same for all elements of $\mathscr{O}$, let us assume the algorithm first considers o_1 as in Example 2.10. As written, we would take the minimum element in the Fibonacci heap (a constant time operation). We would then consider the partners for o_1 which would result in the greedy selection of p_2, (just as in GREEDY-KSEP-OPT1 and NAIVE-KSEP-SC. Also note that we retain the array of Booleans, $\mathcal{O}'$ as well as the array of lists, M to help us with these operations).

Now the issue arises that we must update the keys for the remaining observations, as well as remove observations covered by p_2. As we maintain REL_OBS and $\mathcal{O}'$, the procedure quickly iterates through the elements covered by p_2: o_1 and o_2. Figure 2.8 shows the status of the observations at this point.

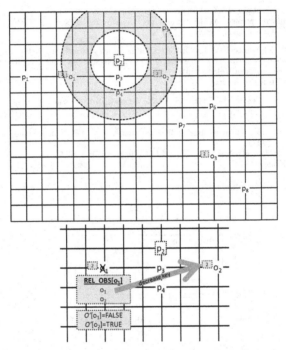

Fig. 2.8 Above: GREEDY-KESP-OPT2 considers all observations that can be partnered with p_2. Notice that in this figure by each observation we show a box that represents the key of the observation in the Fibonacci heap. **Below:** GREEDY-KSEP-OPT2 removes o_1 from the heap, and iterates through the elements in REL_OBS[o_1], causing it to decrease the key of o_2.

We remove o_1 from the heap, and set $\mathcal{O}'[o_1]$ to FALSE. This prevents us from considering it in the future. We now iterate through each o'' in the list pointed to by REL_OBS[o_1] where $\mathcal{O}'[o'']$ is TRUE and decrease the key of each by one. As per Table 2.2, REL_OBS[o_1] = $\{o_1, o_2\}$. As $\mathcal{O}'[o_1]$ = FALSE we do nothing. As $\mathcal{O}'[o_2]$ = TRUE, we decrease the key of the associated node in the Fibonacci heap. The array QUICK_LOOK ensures we can access that element in constant time. Figure 2.8 (left) graphically depicts this action.

Next, we consider the other element covered by partner p_2: o_2. After removing this element from the heap and setting $\mathcal{O}'[o_2]$ to FALSE, we can easily see that there does not exist any $o'' \in$ REL_OBS[o_2] where $\mathcal{O}'[o'']$ = TRUE. Hence, we can proceed to pick a new minimum observation from the heap - which is o_3 in this case. The greedy selection proceeds (resulting in the choice of p_6), followed by the update procedure (which simply removes the node associated with o_3 from the heap and sets $\mathcal{O}'[o_3] = \mathsf{FALSE}$). As there are no more elements in the heap, GREEDY-KSEP-OPT2 returns the solution $\{p_2, p_6\}$.

The result below analyzes the complexity of the GREEDY-KSEP-OPT2 algorithm.

Theorem 2.6 (Complexity of GREEDY-KSEP-OPT2). *GREEDY-KSEP-OPT2 has a complexity of $O(\Delta \cdot f^2 \cdot |\mathcal{O}| + |\mathcal{O}| \cdot \ln(|\mathcal{O}|))$ and an approximation ratio of $1 + \ln(f)$.*

Proof. CLAIM 1: GREEDY-KSEP-OPT2 has a complexity of $O(\Delta \cdot f^2 \cdot |\mathcal{O}| + |\mathcal{O}| \cdot \ln(|\mathcal{O}|))$.

Line 1 takes $O(\Delta \cdot |\mathcal{O}|)$ time.

The loop starting at line 4 iterates $|\mathcal{O}|$ times.

The nested loop at line 4a iterates Δ times.

The second nested loop at line 4(a)i iterates f times. The inner body of this loop can be accomplished in constant time.

In line 5, initializing the Fibonacci heap takes constant time, as does inserting elements, hence this line takes only $O(|\mathcal{O}|)$ time.

The loop at line 6 iterates, at most, $|\mathcal{O}|$ times.

Viewing the minimum of a Fibonacci heap, as in line 6a can be done in constant time.

As per the analysis of GREEDY-KSEP-OPT1, line 6b takes $\Delta \cdot f$ iterations. The updating procedure starts with line 6c which iterates f times.

The removal of an element in line 6(c)ii from a Fibonacci heap costs $O(\ln(|\mathcal{O}|)$ amortized time. However, we perform this operation no more than $|\mathcal{O}|$ times, hence we can add $|\mathcal{O}| \cdot \ln(|\mathcal{O}|))$ to the complexity.

Note that the size of a list pointed to by REL_OBS[o'] is bounded by $\Delta \cdot f$—f observations associated with each of Δ partners—hence line 6(c)iii iterates, at most, $\Delta \cdot f$ times.

We note that decreasing the key of an item in the Fibonacci heap (in line 6(c)iii) takes constant time (amortized).

Therefore, by the above statements, the complexity of GREEDY-KSEP-OPT2 is $O(|\mathcal{O}| \cdot (\Delta \cdot f + \Delta \cdot f^2) + |\mathcal{O}| \cdot \ln(|\mathcal{O}|) + \Delta \cdot f \cdot |\mathcal{O}| + \Delta \cdot |\mathcal{O}|)$ and the statement follows.

CLAIM 2: GREEDY-KSEP-OPT2 has an approximation ratio of $1 + \ln(f)$.

Follows directly from Proposition 2.11.

The result below establishes the soundness of the GREEDY-KSEP-OPT2 algorithm.

Proposition 2.13. *GREEDY-KSEP-OPT2 returns a $|\mathcal{E}|$-sized (α, β) explanation for $\mathcal{O}$.*

GREEDY-KSEP-OPT2 returns IMPOSSIBLE if there is no explanation for $\mathcal{O}$.

Proof. Mirrors that of Proposition 2.12.

2.6 Implementation and Experiments

In this section, we show that our geospatial abduction framework and algorithms are viable in solving real-world geospatial abduction problems. Using a real-world data set consisting of counter-insurgency information from Iraq, we were able to accurately locate insurgent weapons cache sites (partners) given previous attacks (observations) and some additional data (used for feas and α, β). This validates our primary research goal for the experiments - to show that geospatial abduction can be used to solve problems in the real-world.

We considered the naive set-covering approach along with GREEDY-KSEP-OPT1 and GREEDY-KSEP-OPT2, which according to our analytical results had the best approximation ratios and time-complexities. We implemented these algorithms in 4000 lines of Java code, running on a Lenovo T400 ThinkPad laptop running Vista with an Intel Core 2 Duo T9400 2.53 GHz processor and 4.0 GB of RAM.

Our SCARE (Social-Cultural Abductive Reasoning Engine) system [25] enabled us to carry out tests on real-world data. This data includes 21 months of Improvised Explosive Device or IED attacks in Baghdad[4] (a 25x27 km region)—these constitute our observations. It also included information on locations of caches associated with those attacks discovered by US forces. The locations of the caches constitute the (α, β) explanation we want to learn. We used data from the International Medical Corps to define feasibility predicates which took the following factors into account:

- the ethnic makeup of neighborhoods in Baghdad—specifically, Sunni locations were deemed infeasible for cache locations;
- the locations of US bases in Baghdad were also considered infeasible; and
- bodies of water were also deemed infeasible.

We also separately ran tests on that part of the above data focused on Sadr City (a 7x7 km district in Baghdad) alone. On both these regions, we overlaid a grid whose cells were 100m x 100m each—about the size of a standard US city block. All timings were averaged over 100 runs.

We split the data into 2 parts—the first 7 months of data was used as a "training" set and the next 14 months of data was used for experimental evaluation. We used the following simple algorithm, FIND-BOUNDS, to automatically learn the α, β values. We set β_{max} to 2.5 km. While it is possible to develop more advanced procedures for learning these parameters, this is not the focus of this book. Such parameters could also come from an expert.

Accuracy. Our primary goal in the experiments was to determine if the geospatial abduction framework and algorithms could provide highly accurate results in a real-world setting. "Accuracy" in this section refers to two aspects: size of the solution and the distance to the nearest actual cache site. The distance to nearest cache site was measured by taking the straight-line Euclidean distance to the nearest cache site that was found after the first attack supported by the projected cache site. We used

---

[4] Attack and cache location data was provided by the Institute for the Study of War.

Algorithm 9 (FIND-BOUNDS)

INPUT: Historical, time-stamped observations $\mathcal{O}_h$, historical, time-stamped partners, $\mathcal{E}_h$, real number (distance threshold) β_{max}
OUTPUT: Real numbers α, β

1. Set $\alpha = 0$ and $\beta = \beta_{max}$
2. Set Boolean variable $flag$ to TRUE
3. For each $o \in \mathcal{O}_h$, do the following:

    a. For each $p \in \mathcal{E}_h$ that occurs after o, do the following.
       i. Let d be the Euclidean distance function.
      ii. If $flag$, and $d(o,p) \leq \beta_{max}$ then set $\alpha = d(o,p)$ and $\beta = d(o,p)$
      iii. If not $flag$, then do the following:
         A. If $d(o,p) < \alpha$ then set $\alpha = d(o,p)$
         B. If $d(o,p) > \beta$ and $d(o,p) \leq \beta_{max}$ then set $\beta = d(o,p)$

4. Return reals α, β

the raw coordinate for the actual cache in the data set—not the position closest to the nearest point in the 100 m resolution grid that we overlaid on the areas. The accuracy results are summarized in Tables 2.3-2.4.

Area	Algorithm	Sample Mean Solution Size	Sample Mean Number of Partners ≤ 0.5 km to actual cache
Baghdad	NAIVE-KSEP-SC	14.53	8.13
	GREEDY-KSEP-OPT1	15.02	7.89
	GREEDY-KSEP-OPT2	14.00	7.49
Sadr City	NAIVE-KSEP-SC	8.00	3.00
	GREEDY-KSEP-OPT1	6.61	4.44
	GREEDY-KSEP-OPT2	6.00	5.28

Table 2.3 k-**SEP** Algorithm Results - Solution Size

Area	Algorithm	Sample Mean Avg Dist to actual cache	Sample Std Dev of Avg Dist to actual cache	Sample Mean Std Dev of Dist to actual cache
Baghdad	NAIVE-KSEP-SC	0.79 km	0.02	0.64
	GREEDY-KSEP-OPT1	0.76 km	0.07	0.60
	GREEDY-KSEP-OPT2	0.72 km	0.03	0.63
Sadr City	NAIVE-KSEP-SC	0.72 km	0.03	0.46
	GREEDY-KSEP-OPT1	0.45 km	0.03	0.46
	GREEDY-KSEP-OPT2	0.35 km	0.03	0.47

Table 2.4 k-**SEP** Algorithm Results - Distances to Actual Cache Sites

Overall, GREEDY-KSEP-OPT2 consistently found the smallest solution—of cardinality 14 for Baghdad and 6 for Sadr City—on all 100 trials. For Baghdad, the other two algorithms both found a solution of size 14, but both averaged a higher cost solution. For Sadr City, GREEDY-KSEP-OPT1 often did find a solution of 6 caches while NAIVE-KSEP-SC only found solutions of size 8. Additionally, in both tests, the solution sizes for GREEDY-KSEP-OPT1 varied more than the other two algorithms. Moreover, Tukey's Honest Significant Difference (HSD) test for both Baghdad and Sadr City indicated significant difference between all pairs of algorithms with respect to solution size.

Of the partners in a given solution, we also recorded the number of partners less than 0.5km away from an actual cache. For Baghdad, NAIVE-KSEP-SC performed best in this regard, averaging 8.13 partners less than 0.5km from an actual cache site. Although this result for Baghdad is significant based on an analysis of variance (ANOVA) and honest significant differences (HSD) (p-value of $2.3 \cdot 10^{-9}$), we also note that the greatest difference among averages was still less than one partner. This same result for Sadr City, however, tells a different story. For this test, NAIVE-KSEP-SC performed poorly with regard to the other two algorithms, only finding 3 partners meeting these criteria for each of the 100 trials. GREEDY-KSEP-OPT2 performed very well in this aspect (for Sadr City). It averaged over 5 partners less than 0.5km from an actual cache. Furthermore, for Sadr City, *all* partners found by GREEDY-KSEP-OPT2 were within 600m of an actual cache site. The analysis of variance, or ANOVA, (p-value of $2.2 \cdot 10^{-16}$) and HSD of partners less than 0.5km from an actual cache for the Sadr City trials indicate that these results are significant.

Our primary metric of accuracy was average distance to an actual cache. In this regard, GREEDY-KSEP-OPT2 performed the best. It obtained an average distance of 0.72km for Baghdad and 0.35km for Sadr City. This number was 40m less for Baghdad and 100 m less for Sadr City when compared to GREEDY-KSEP-OPT1, whose average distance varied widely among the trials. With regard to this metric, NAIVE-KSEP-SC performed the worst—particularly in Sadr City, where it predicted caches over twice as far from actual caches as GREEDY-KSEP-OPT2 (on average). For both Baghdad and Sadr City, the simple ANOVA yielded a p-value of $2.2 \cdot 10^{-16}$, which suggests with over a 99% probability that there is a difference among the algorithms. Also, for both areas, Tukey's HSD indicates significant difference between each pair-wise comparison of algorithms.

Algorithm run times. Table 2.5 shows the run times of our algorithms. In order to validate the findings suggested by Table 2.5 statistically, we ran analysis of variance (ANOVA) and Tukey's Honest Significant Difference test (HSD) for pairwise comparisons [8]. An ANOVA for the Baghdad run times gave a p-value of $2.2 \cdot 10^{-16}$, which suggests with well over 99% probability that GREEDY-KSEP-OPT1 is statistically faster than GREEDY-KSEP-OPT2. The HSD for Baghdad indicates that, with regard to run times, all pair-wise-comparison of the three algorithms are significantly different. For Sadr City, the ANOVA gave a p-value of $4.9 \cdot 10^{-3}$, which suggests with a 99% probability that the algorithms differ in runtimes. However, the HSD indicates, with an 82% probability, that there is no difference among GREEDY-KSEP-OPT1 and GREEDY-KSEP-OPT2, while both differ significantly from NAIVE-KSEP-SC.

Area	Algorithm	Sample Mean Run-Time	Sample Run-Time Standard Deviation
Baghdad	NAIVE-KSEP-SC	354.75 ms	12.86
	GREEDY-KSEP-OPT1	162.08 ms	40.83
	GREEDY-KSEP-OPT2	201.40 ms	36.44
Sadr City	NAIVE-KSEP-SC	28.85 ms	10.52
	GREEDY-KSEP-OPT1	25.44 ms	9.33
	GREEDY-KSEP-OPT2	24.64 ms	8.95

Table 2.5 *k*-**SEP** algorithm performance results.

2.6.1 A Simple Heuristic to Improve Accuracy

In our implementation of all three algorithms, ties in greedy selection of partners were determined by a "coin toss." Specifically, we are considering the case where $this\_size = cur\_size$ in line 4(b)iii of **NAIVE-KSEP-SC** in Section 2.4.2. Let us rephrase the situation as follows. Let $\mathcal{O}$ be the entire set of observations and $\mathcal{O}' \subseteq \mathcal{O}$ be the set of observations currently not assigned a partner. Let p be the current partner that best meets the criteria for greedy selection and p' be the partner we are considering. We define P and P' as subsets of $\mathcal{O}$ that are the observations associated with p and p' respectively. Hence, if $|P' \cap \mathcal{O}'| > |P \cap \mathcal{O}'|$, we pick p'. As implemented, if $|P' \cap \mathcal{O}'| = |P \cap \mathcal{O}'|$, we flip a coin. We add a simple heuristic that simply states that partners that cover more observations are preferred. We change the criteria as follows:

- If $|P' \cap \mathcal{O}'| = |P \cap \mathcal{O}'|$, then do the following:

  - If $|P'| > |P|$, pick p'
  - If $|P| > |P'|$, pick p
  - If $|P| = |P'|$, flip a coin

We refer to this as the "tie-breaker" heuristic. The result is that the solution set of partners covers more observations and hence provides a more "dense" solution.

We added this heuristic to our existing code for all three algorithms and ran each one 100 times for both the Baghdad and Sadr City areas. Unsurprisingly, as this is a constant-time operation, run-times were not affected. However, accuracy improved in all cases. As GREEDY-KSEP-OPT2 still provided the most accurate results, the following exposition focuses on how the heuristics affected the solution size and accuracy for this algorithm.

Because the tie-breaker heuristic only adjusts how two partners are chosen—both of which can be paired with the same uncovered observations—the size of

Area	Tie-Breaker Heuristic	Sample Mean Solution Size	Sample Mean Number of Partners ≤ 0.5 km to actual cache
Baghdad	No	14.00	7.49
	Yes	14.00	7.87
Sadr City	No	6.00	5.28
	Yes	6.00	6.00

Table 2.6 The Tie-Breaker heuristic on GREEDY-KSEP-OPT2: solution size

Area	Tie-Breaker Heuristic	Sample Mean Avg Dist to actual cache	Sample Std Dev of Avg Dist to actual cache	Sample Mean Std Dev of Dist to actual cache
Baghdad	No	0.72 km	0.03	0.63
	Yes	0.69 km	0.02	0.64
Sadr City	No	0.35 km	0.03	0.47
	Yes	0.28 km	0.02	0.11

Table 2.7 The Tie-Breaker heuristic on GREEDY-KSEP-OPT2: distances to actual cache sites

the solution was unaffected in both the Baghdad and Sadr City trials. However, the number of predicted cache sites less than 500m from an actual site increased for both the Baghdad and Sadr City tests. For Baghdad, more trials returned solutions with 8 predictions less than 500m from an actual site than returned 7—the opposite being the case without the tie-breaker heuristic. For Sadr City, all elements of every solution set returned were less than 500m from an actual cache site. Using the well known T-Test [8], we showed that these results are statistically significant as this test returned a p-value of $6.2 \cdot 10^{-8}$ for Baghdad and $2.2 \cdot 10^{-16}$ for Sadr City.

Summary. The above experiments demonstrate statistically that GREEDY-KSEP-OPT2 provides a viable solution—consistently producing smaller solution sets which were closer to actual cache sites faster than the basic set-covering approach, at times approaching the faster, although less-accurate GREEDY-KSEP-OPT1. The proximity of the elements of the solution set to actual cache sites is encouraging for real-world use. The results are strong enough that two US Army units used SCARE to aid in locating IED caches.

References

1. Alpaydin, E.: 2010. *Introduction to Machine Learning*. MIT Press, 2 edition, 2010.
2. Brantingham, P., Brantingham, P.: 2008. Crime Pattern Theory. In Enviromental Criminology and Crime Analysis, R. Wortley and L. Mazerolle, Eds., pages 78-93.
3. Cormen, T.H., Leiserson, C.E., Rivest, R.L., Stein, C.: 2001. Introduction to Algorithms. MIT Press, second edition, 2001.

4. Hunt, III, H. B., Marathe, M. V., Radhakrishnan, V., Stearns, R. E. 1998. The complexity of planar counting problems. *SIAM J. Comput. 27*, 4, 1142–1167.

5. S. Masuyama, T. Ibaraki, T. H. 1981. The computational complexity of the m-center problems on the plane. *Trans. IECE of Japan E84*, 57–64.

6. Dyer, M., Goldberg, L. A., Greenhill, C., Jerrum, M. 2000. On the relative complexity of approximate counting problems. Tech. rep., Coventry, UK, UK.

7. Fredman, M.L., Tarjan, R.E.: 1987. Fibonacci heaps and their uses in improved network optimization algorithms. Journal of the ACM, 34(3):596–615, July 1987.

8. Freedman, D., Purves, R., Pisani, R.: 2007. Statistics. W.W. Norton and Co., 4 edition.

9. Garey, M.R., Johnson, D.S.: 1979. Computers and Intractability; A Guide to the Theory of NP-Completeness. W. H. Freeman & Co., New York, NY, USA.

10. Hochbaum, D.S.: 1982. Approximation Algorithms for the Set Covering and Vertex Cover Problems. SIAM Journal on Computing, 11(3):555–556.

11. Hochbaum, D.S.: 1997. *Approximation Algorithms for NP-Complete Problems*. PWS Publishing Co., 1997.

12. Hochbaum, D.S., Maass, W.: 1985. Approximation schemes for covering and packing problems in image processing and vlsi. Journal of the ACM, 32:130–136.

13. Jia, L., Rajaraman, R. Suel, T.: 2002. An efficient distributed algorithm for constructing small dominating sets. Distrib. Comput., 15(4):193–205.

14. Johnson, D.S.: 1982. The np-completeness column: An ongoing guide. Journal of Algorithms, 3(2):182–195, 1982.

15. Karp, R.: 1972. Reducibility Among Combinatorial Problems. In R. E. Miller and J. W. Thatcher, editors, Complexity of Computer Computations, page 85-103.

16. Kuhn, F., Wattenhofer,R.: 2003. Constant-time distributed dominating set approximation. In In Proc. of the 22 nd ACM Symposium on the Principles of Distributed Computing (PODC, pages 25–32.

17. Lu, J., Nerode, A., Subrahmanian, V.S.: 1996. Hybrid Knowledge Bases, IEEE Transactions on Knowledge and Data Engineering, 8, 5, pages 773-785.

18. Lund, C., Yannakakis, M.: 1994. On the hardness of approximating minimization problems. Journal of the ACM, 41(5):960–981.

19. Papadimitriou, C.H.: 1981. Worst-Case and Probabilistic Analysis of a Geometric Location Problem, *SIAM J. Comput.*, 10(3):542–557.

20. Paschos, V.T. 1997.: A survey of approximately optimal solutions to some covering and packing problems. ACM Comput. Surv., 29(2):171–209.

21. Reggia, J.A., Peng, Y.: 1990. Abductive inference models for diagnostic problem-solving. Springer-Verlag New York, Inc., New York, NY, USA.

22. Rimoin, A. *et al.*: Endemic Human Monkeypox, Democratic Republic of Congo, 2001-2004, Emerging Infectious Diseases, 13, 6, pages 934–937, 2007.

23. Rossmo, D. K., Rombouts, S.: 2008. Geographic Profiling. In Enviromental Criminology and Crime Analysis, R. Wortley and L. Mazerolle, Eds. pages 136-149.

24. H. Samet.: The Design and Analysis of Spatial Data Structures, Addison Wesley, 1989.

25. Shakarian, P., Subrahmanian, V.S., Sapino, M.L. SCARE: A Case Study with Baghdad, Proc. 2009 Intl. Conf. on Computational Cultural Dynamics (eds. D. Nau, A. Mannes), Dec. 2009, AAAI Press.

26. Shakarian, P., Subrahmanian, V.S., Sapino, M.L. 2012. GAPS: Geospatial Abduction Problems, ACM Transactions on Intelligent Systems and Technology (TIST), 3, 1, to appear.

27. Shakarian, P., Subrahmanian, V.S. Region-based Geospatial Abduction with Counter-IED Applications, accepted for publication in: Wiil, U.K. (ed.).Counterterrorism and Open Source Intelligence, Springer Verlag Lecture Notes on Social Networks, to appear, 2011.

28. Shakarian, P., Nagel, M., Schuetzle, B., Subrahmanian, V.S. 2011. Abductive Inference for Combat: Using SCARE-S2 to Find High-Value Targets in Afghanistan, in Proc. 2011 Intl. Conf. on Innovative Applications of Artificial Intelligence, Aug. 2011, AAAI Press.

29. Shakarian, P., Dickerson, J., Subrahmanian, V.S. 2012. Adversarial Geospatial Abduction Problems, ACM Transactions on Intelligent Systems and Technology (TIST), to appear.

30. Singh, M., Joshi, P.K., Kumar,M., Dash, P.P., Joshi, B.D.: Development of tiger habitat suitability model using geospatial tools: a case study in Achankmar Wildlife Sanctuary (AMWLS), Chhattisgarh India, Env. Monitoring and Assessment journal, Vol. 155, pages 555-567, 2009.
31. US Army: *Intelligence Preparation of the Battlefiled (US Army Field Manual)*, FM 34-130 edition, 1994.
32. Vazirani, V.V.: 2004. Approximation Algorithms. Springer, March 2004.

Chapter 3
Region-Based Geospatial Abduction

Abstract Given a set $\mathcal{O}$ of observations, in the previous chapter, we developed a set of methods to find sets $\mathcal{E}$ of explanations. However, these explanations consisted of points. When the geospatial resolution of the space $\mathcal{S}$ is small, the non-determinism in our point-based geospatial abduction algorithms is negligible—making many points more or less "equivalent" as far as being potential partner locations. As such, users might want to get regions back as output to their geospatial abduction queries. Moreover, users might want to reason about real-valued points rather than points that are integer-valued. In this chapter, we develop the theory and algorithms required for reasoning in the real-valued domain with regions being returned to the user rather than points.

3.1 Introduction

In Chapter 2, we developed a theory of geospatial abduction in which a set of points was returned as an answer (or explanation) to the user. For instance, in our IED cache detection problem, we returned a set of points consisting of potential locations of IED caches in Baghdad. In our tiger detection problem, we returned sets of points where a tiger might dwell, given the locations of various kills attributed to the tiger. In the same vein, in our virus host detection problem, we returned sets of points where the host of a virus causing a disease such as monkey-pox might conceivably reside. And in our burglar detection problem, the explanations generated by point-based geospatial abduction identified explanations consisting of points where a burglar might reside (*e.g.*, his house, his office, his significant other's house, etc.).

However, when our space $\mathcal{S}$ has a fine-grained resolution, many points might be more or less "equivalent" as far as being potential IED cache locations is concerned. As a consequence, point-based geospatial abduction yields many potential points that could be included in an explanation, and sometimes, preferring one point to another is merely a matter of non-deterministic choice, rather than rational preference of one point over another.

In this chapter, we try to return explanations for geospatial abduction that consist of sets of *regions* rather than points so that such non-determinism can be significantly reduced. Thus, our definition of an explanation in this chapter returns a set or regions. Each region in an explanation says a potential IED weapons cache (or a tiger dwelling, or a region supporting a virus host, or a set of locations corresponding to a burglar's residence) might be somewhere (*i.e.*, at any point) within the region.

In addition, in this chapter, we focus on the real-valued domain. While most real-world GIS systems only use integer-valued coordinates, real-valued coordinates are interesting both from a theoretical perspective and often from the perspective of better computation—for example, solving linear constraints over the continuous, real-valued domain is polynomial, while solving the same linear constraints over the domain of the integers is well known to be NP-hard.

3.2 Technical Preliminaries

To address the problem of region-based geospatial abduction, we introduce a framework that resembles that of Chapter 2—but differs in several important aspects. These include the use of a continuous space and multiple types of explanations. In Chapter 4, we return to the original framework of Chapter 2.

Unlike the previous chapter, we assume the existence of a real-valued $M \times N$ space $\mathscr{S}$ whose elements are pairs of *real* numbers (rather than integers) from the set $[0,M] \times [0,N]$. An observation is any member of $\mathscr{S}$—thus, unlike the preceding chapter, observations are pairs of real values. We use $\mathscr{O}$ to denote an arbitrary, but fixed, finite set of *observations*. We assume there are real numbers $\alpha \leq \beta$ such that for each observation o, there exists a partner p_o (to be found) whose distance from o is in the interval $[\alpha, \beta]$.[1] Without loss of generality, we also assume that all elements of $\mathscr{O}$ are over β distance away from the edge of $\mathscr{S}$. Example 3.1 presents a neighborhood as a space and locations tiger dwellings.

Example 3.1 (Tiger Example). A tiger in the Achanakamar Wildlife Sanctuary (AMWLS) has made many kills. Suppose the AMWLS sanctuary is the space $\mathscr{S}$ depicted in Figure 3.1. Tiger kills were found by wildlife rangers at points $\mathscr{O} = \{o_1, \ldots, o_{13}\}$. Tiger conservation experts, on the basis of historical data, suggest that favored tiger dwellings are located within 5km of these kills (*i.e.*, $\alpha = 0$ and $\beta = 5$km). Note that in Figure 3.1, circles of radius 5km are drawn around the observation points. The tiger conservation experts are interested in the locations of such dwellings.

Throughout this chapter, we assume the notion of a *distance function d* on $\mathscr{S}$ satisfying the usual properties of such distance functions introduced in Chapter 2. The methods used in this chapter apply to *any notion of distance between two points as long as the three distance axioms described in Chapter 2 are satisfied.*

---

[1] Chapter 2 describes methods to learn α, β automatically from historical data.

We now define a region and how they relate to the set of observations. Our intuition is simple—a region *explains* an observation if that region contains a partner point for that observation.

Definition 3.1 (Region / Super-Explanation / Sub-Explanation). A *region r* is a subset of $\mathscr{S}$ such that for any two points $(x,y),(x',y') \in r$, there is sequence a of line segments from (x,y) to (x',y') s.t. no line segment lies outside r.

1. A region r **super-explains** point o in $\mathscr{S}$ iff there exists a point $p \in r$ such that $d(o,p) \in [\alpha,\beta]$.
2. A region r **sub-explains** some point o in $\mathscr{S}$ iff $(\forall p \in r)\, d(o,p) \in [\alpha,\beta]$.

Thus, intuitively, a region r as defined above is connected in the sense that one can travel from any point in a region to any other point in the region without leaving the region r. In addition, regions can have any shape and may overlap.

Informally speaking, region r super-explains an observation o if and only if there is at least one partner in region r for the observation o. On the other hand, region r sub-explains an observation o if and only if every point in the region explains observation o. Throughout this chapter, we assume that checking if some point o is sub-explained (super-explained) by region r can be performed in constant (*i.e.*, $O(1)$) time. This is a reasonable assumption for most regular shaped regions like circles, ellipses and polygons. The following result follows immediately from Definition 3.1.

Observation 3.2.1 *If region $r \neq \emptyset$ sub-explains point o, then r super-explains point o.*

This observation follows immediately from the definitions. If r sub-explains point o then the distance of every point in r from observation o lies within the interval $[\alpha,\beta]$. Thus, as long as r is non-empty, at least one point in r is at a distance d_0 from the observation o where $d_o \in [\alpha,\beta]$.

We would like to explain observations by finding regions containing a partner. In some applications, the user may be able to easily search the entire region—hence a super-explaining region would suffice. In other applications, we may want to be sure that any point within the region can be a partner as not to waste resources—so only a sub-explanation would make sense in such a case. Often, these situations may depend on the size of the regions. We shall discuss the issue of restricting region size later in this section. For now, we shall consider regions of any shape or size. Example 3.2 shows regions that super- or sub-explain various observations.

Example 3.2. Consider the scenario from Example 3.1 and the regions $R = \{r_a, r_b, r_c, r_d, r_e, r_f, r_g\}$ shown in Figure 3.1. Suppose these regions correspond with feasible regions for the tiger to live in—*i.e.*, places that have the right amount of ground cover and the right amount of prey for a tiger to consider this to be a good habitat. Consider region r_a. As it totally lies within the α,β distance of o_1, it both sub-explains and super-explains this observation. Conversely, region r_d super-explains both o_6 and o_7 but sub-explains neither.

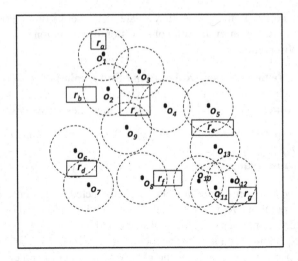

Fig. 3.1 Locations of tiger kills and feasible locations $\{r_a, r_b, r_c, r_d, r_e, r_f, r_g\}$ where the tiger can potentially dwell. The β distance for each observation is shown with a dashed circle.

This chapter studies following decision problems.

Sub-(Super-) Region Explanation Problem (Sub/Sup-REP)
INPUT: A space $\mathscr{S}$, distance interval $[\alpha, \beta]$, set $\mathscr{O}$ of observations, set R of regions, and natural number $k \in [1, |\mathscr{O}|]$.
OUTPUT: Set $R' \subseteq R$, where $|R'| \leq k$ and for each $o \in \mathscr{O}$, there is an $r \in R$ such that r sub-(super-) explains o.

The Sub-(Super) Region Explanation Problem asks us to find all sub-explanations (resp. super-explanations) R of size k or fewer where size is defined as the number of regions in R which sub-explain (resp. or super-explain) our set of observations $\mathscr{O}$.

The fact that a set R of regions is part of the input is not an assumption, but a feature. A user might set R to be all the regions associated with $\mathscr{S}$ in which case he is really making no assumption at all. Alternatively, he might use his knowledge of the application (*e.g.*, IED cache locations or tiger hangouts or virus host information or burglary-related information) to define regions, taking into account, the terrain and/or known aspects of the population living in the area of interest. For instance, when trying to identify regions containing IED caches in Baghdad used for attacks by Shi'ite groups, he might define regions to be places that are not predominantly Sunni and that do not contain US bases or bodies of water. On the other hand, in the tiger detection application, he might define regions to be places where the tiger has ample ground cover and ample amount of prey to hunt. In the virus host detection problem, he might decide based on his knowledge of biology and his knowledge of the geography of the terrain, that certain regions are feasible locations for the virus host, while others are not. And finally, the St. Paul, MN, police detective might use knowledge of the criminal to decide that the criminal could not live in certain areas (*e.g.*, there was a police chase not known to the geospatial abduction system where the perpetrator disappear in a reasonably narrow region, allowing the detective to eliminate other regions from consideration). Other kinds of logical conditions may be used when dealing with burglaries or drug trafficking.

Thus, the set R of regions allows an analyst to specify any knowledge he has, and allows the system to benefit from that knowledge. In short, the set R is similar to the feasibility predicate in Chapter 2 by saying that only regions in R can be returned as part of the answer by the region-based geospatial abduction system. If no such knowledge is available, R can be taken to be the set of all regions associated with $\mathscr{S}$, and thus, allowing the user to specify R as part of the input leads to no loss of generality; moreover, it allows the user greater flexibility in specifying where the regions he is looking for could possibly be. R can also be used to restrict the size of the region (*e.g.*, only considering regions whose area is less than 5 sq. km.).

There are two different associated optimization problems associated with both the Sub-REP and Sup-REP problems. The first deals with finding a subset of regions of minimal cardinality that explains all observations.

Sub-(Super-)Region Explanation Problem-Minimum Cardinality (Sub/Sup-REP-MC)

INPUT: A space, $\mathscr{S}$, distance interval $[\alpha, \beta]$, set of observations $\mathscr{O}$, and set of regions R.
OUTPUT: Set $R' \subseteq R$ of minimum cardinality, where for each $o \in \mathscr{O}$, there is an $r \in R$ s.t. r sub-(super-) explains o.

The Sub/Sup-REP-MC problems therefore support the principles of Occam's razor, long present in research on abduction[3, 6]. Only a minimal-sized set of regions can be returned—no more regions than strictly necessary should be returned.

Our second optimization problem fixes the number of regions returned in the solution, but maximizes the number of observations that are explained.

Sub-(Super-)Region Explanation Problem-Maximum Explaining (Sub/Sup-REP-ME)

INPUT: Given a space $\mathscr{S}$, distance interval $[\alpha, \beta]$, set $\mathscr{O}$ of observations, set R of regions, and natural number $k \in [1, |\mathscr{O}|]$.
OUTPUT: Set $R' \subseteq R$, where $|R'| \leq k$ such that the number of $o \in \mathscr{O}$ where there is an $r \in R$ s.t. r sub-(super-) explains o is maximized.

Sub-(Super-)Region Explanation Problem-Maximum Explaining (Sub/Sup-REP-ME) problems are similar in spirit to the k-**SEP** problem by requiring that no more than k regions be returned as the answer by the geospatial abduction system in response to a user request. Consider the following example.

Example 3.3. Consider the scenario from Example 3.2. Consider an instance of Sup-REP with $k = 7$. The set $\{r_a, r_b, r_c, r_d, r_e, r_f, r_g\}$ is a solution to this problem. Now consider Sup-REP-MC: the set $\{r_a, r_c, r_d, r_e, r_f, r_g\}$ is a solution to this problem. Finally, consider Sup-REP-ME with $k = 2$. The set $\{r_c, r_d\}$ is a solution to this problem.

We now consider a special case of these problems that arises when the set R of regions is created by a partition of the space based on the set of observations $(\mathcal{O})$ and concentric circles of radii α and β drawn around each $o \in \mathcal{O}$. We can associate regions in such a case with subsets of $\mathcal{O}$. For a given subset $\mathcal{O}'$, we say that there is an associated set of *induced regions* (denoted $R_{\mathcal{O}'}$), defined as follows:

$$R_{\mathcal{O}'} = \{\{x| \ \forall o \in \mathcal{O}', d(x,o) \in [\alpha, \beta] \land$$
$$\forall o' \notin \mathcal{O}', d(x,o') \notin [\alpha, \beta]\} \ \}$$

We note that for a given subset of observations, it is possible to have a set of induced regions, $R_{\mathcal{O}'}$ that has more than one element. For example, consider set $R_\emptyset = \{r_1, r_{12}\}$ in Figure 3.2. For a given set of observations $\mathcal{O}$, we will use the notation $R_{\mathcal{O}}$ do denote the set of all induced regions. Formally:

$$R_{\mathcal{O}} = \bigcup_{\substack{\mathcal{O}' \in 2^{\mathcal{O}} \\ R_{\mathcal{O}'} \neq \emptyset}} R_{\mathcal{O}'}$$

We illustrate the idea of induced regions in the following example.

Example 3.4. In order to identify where the tiger resides, tiger conservation experts may create 33 **induced** regions in $\mathscr{S}$ by drawing circles of 5km radius around all observations (see Figure 3.2), the set of which is denoted $R_{\mathcal{O}} = \{r_1, \ldots, r_{33}\}$.

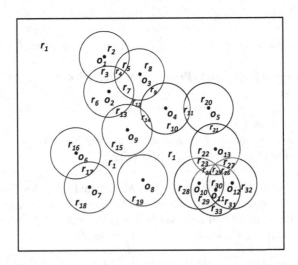

Fig. 3.2 Space $\mathscr{S}$ and the regions in set $R_{\mathcal{O}}$.

For the special case where $R_{\mathcal{O}}$ is the set of all possible regions of $\mathcal{S}$, we have the following result.

Lemma 3.1. *Suppose $\mathcal{O}$ is a set of observations and $R_{\mathcal{O}}$ is the set of induced regions. A region $r \in R_{\mathcal{O}}$ sub-explains an observation $o \in \mathcal{O}$ if and only if it super-explains o.*

Proof. CLAIM 1: Any point in a region $r \in R_{\mathcal{O}}$ is either within distance $[\alpha, \beta]$ or outside the distance $[\alpha, \beta]$ from each $o \in \mathcal{O}$.
As $R_{\mathcal{O}}$ is created by drawing circles of radii α, β around each observation, the statement follows by the definition of $R_{\mathcal{O}}$.

CLAIM 2: ($\Leftarrow$) There is no $r \in R_{\mathcal{O}}$ that super-explains some $o \in \mathcal{O}$ but does not sub-explain the observation.
Suppose, by way of contradiction, there is some $r \in R_{\mathcal{O}}$ that super-explains some $o \in \mathcal{O}$ but does not sub-explain it. Then, there must be at least one point in r that can be partnered with $\mathcal{O}$ and at least one point in r that cannot be partnered with o. However, by Claim 1, this is not possible, hence a contradiction.
CLAIM 3: ($\Rightarrow$) There is no $r \in R_{\mathcal{O}}$ that sub-explains some $o \in \mathcal{O}$ but does not super-explain the observation.
Follows directly from Observation 3.2.1.

By this result, for the special case of induced regions, we only need one decision problem.

Induced Region Explanation Problem (I-REP)
INPUT: Given a space, $\mathcal{S}$, distance interval $[\alpha, \beta]$, set $\mathcal{O}$ of observations, and natural number $k \in [1, |\mathcal{O}|]$.
OUTPUT: Set $R' \subseteq R_{\mathcal{O}}$, where $|R'| \leq k$ and for each $o \in \mathcal{O}$, there is an $r \in R$ s.t. r sub-explains o.

As mentioned earlier, the sizes of regions can be regulated by our choice of R. However, we may also explicitly require that all regions must be less than a certain area. Consider the following variant of Sup-REP.

Area-Constrained Super-Region Explanation Problem (AC-Sup-REP)
INPUT: Given a space, $\mathcal{S}$, distance interval $[\alpha, \beta]$, set $\mathcal{O}$ of observations, set R of regions, area A, and natural number $k \in [1, |\mathcal{O}|]$.
OUTPUT: Set $R' \subseteq R$, where $|R'| \leq k$ and each $r \in R'$ has an area $\leq A$ and for each $o \in \mathcal{O}$, there is an $r \in R$ such that r super-explains o.

The following proposition tells us that AC-Sup-REP is at least as hard as I-REP, yet no harder than Sup-REP (an analogous result can easily be shown for an area-constrained version of Sub-REP). We note that essentially, we eliminate the regions whose area is above area A, which gives us an instance of Sup-REP. To go the other direction, we directly encode I-REP into an instance of AC-Sup-REP and have A be larger than the area of any region.

Theorem 3.1. *I-REP is polynomially reducible to AC-Sup-REP.*
AC-Sup-REP is polynomially reducible to Sup-REP.

Proof. CLAIM 1: I-REP $\leq_p$ AC-Sup-REP.
Set up an instance of AC-Sup-REP with the input for I-REP plus the parameter
$A = \pi \cdot (\beta^2 - \alpha^2)$. For direction $\Leftarrow$, note that a solution to this instance of I-REP is
also a solution to AC-Sup-REP, as any region that sub-explain an observation also
super-explains it for the set of region $R_\mathcal{O}$ (Lemma 3.1) and the fact that, by definition, all regions in the set $R_\mathcal{O}$ must have an area less than A. For direction $\Rightarrow$, we
know that only regions that can be partnered with observations are considered by
the area restriction, and by Lemma 3.1, all regions in the solution are also super-
explanations for their corresponding observation.

CLAIM 2: AC-Sup-REP $\leq_p$ Sup-REP.
Consider the set R from AC-Sup-REP and let set $R' = \{r \in R|$ *the area of* $r \leq A\}$.
Set up an instance of Sup-REP where the set of regions is R' and the rest is the input
from AC-Sup-REP. For direction $\Leftarrow$, it is obvious that any solution to AC-Sup-REP
is also a solution to Sup-REP, as $R - R'$ are all regions that cannot possibly be in the
solution to the instance of AC-Sup-REP. Going the other direction ($\Rightarrow$), we observe
that by the definition of R', all regions in the result of the instance of Sup-REP meet
all the requirements of the AC-Sup-REP problem.

In the final observation of this section, we note that the set $R_\mathcal{O}$ can be used as a
"starting point" in determining regions. For instance, supplemental information on
areas that may be restricted from being partnered with an observation may also be
considered and reduce the area of (or eliminate altogether) some regions in the set.
Consider the following example.

Example 3.5. Consider the tiger scenario from Example 3.4. Tiger conservation experts may eliminate an open meadow in the area and certain other areas with small
amounts of prey from their search. These "restricted areas" are depicted in Figure 3.3. Note that several regions from Figure 3.2 are either eliminated or have
decreased in size. However, by eliminating these areas, tiger conservation experts
have also pruned some possibilities from their search. For example, regions r_9, r_{13}
were totally eliminated from consideration.

3.3 Complexity

In this section, we study the computational complexity of problems related to
region-based geospatial abduction. In particular, we show that Sub-REP, Sup-REP,
and I-REP are NP-Complete and that the associated optimization problems are NP-
Hard. We also show that the optimization problems Sub-REP-MC, Sup-REP-MC,
and I-REP-MC cannot be approximated by a fully polynomial-time approximation scheme (FPTAS) unless $P = NP$. In particular, this means that there are no

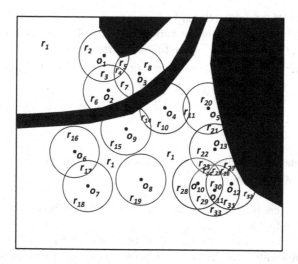

Fig. 3.3 A set of regions in $\mathscr{S}$ created based on the distance $\beta = 5$km as well as restricted areas (shown in black).

polynomial-time algorithms to approximate these problems with guarantees of approximation unless $P = NP$ (the latter, of course, is a central unsolved problem in computer science and it is widely believed that in fact $P \neq NP$). We also note that the complexity of the area-constrained versions of these problems follows directly from the results of this section by the reduction of Theorem 3.1 (page 64).

We first prove that I-REP is NP-complete, which then allows us to correctly identify the complexity classes of the other problems by leveraging Lemma 3.1. First, we introduce the problem of "circle covering" (CC) that was proven to be NP-complete in [11].

Circle Covering (CC)
INPUT: A space $\mathscr{S}'$, set P of points, real number β', natural number k'.
OUTPUT: "Yes" if there is a set of points, Q in $\mathscr{S}'$ such that all points in P are covered by discs centered on points in Q of radius β' where $|Q| \leq k'$—"no" otherwise.

The theorem below establishes that I-REP is NP-complete.

Theorem 3.2. *I-REP is NP-Complete.*

Proof. CLAIM 1: I-REP is in-NP.
Given a set of regions, $R' \subseteq R_{\mathscr{O}}$ we can easily check in polynomial time that for each $o \in \mathscr{O}$ there is an $r \in R$ that is a partner for o. Simply check if each r falls within the distance $[\alpha, \beta]$ for a given $o \in \mathscr{O}$. The operation will take time $O(|\mathscr{O}| \cdot |R'|)$—which is polynomial.

CLAIM 2: I-REP is strongly NP-hard.

We show that for an instance of the known strongly NP-complete problem, circle covering (CC), $CC \leq_p I-REP$ by the following transformation.

- Set $\mathscr{S} = \mathscr{S}'$
- Set $\mathscr{O} = P$
- Set $\beta = \beta'$
- Set $\alpha = 0$
- Set $k = k'$

This transformation obviously takes polynomial time. We prove correctness with the following two sub-claims.

CLAIM 2.1: If there is a k-sized solution R' for I-REP, then there is a corresponding k'-sized solution for CC.

Consider some $r \in R'$. Let $\mathscr{O}'$ be the subset of $\mathscr{O}$ (also of P) such that all points in $\mathscr{O}'$ are partnered with r. By definition, all points enclosed by r are of distance β or less away from each point in $\mathscr{O}'$. Hence, we can pick some point enclosed by r and we have the center of a circle that covers all elements in $\mathscr{O}'$. The statement follows.

CLAIM 2.2: If there is a k'-sized solution Q for CC, then there is a corresponding k-sized set solution for I-REP.

Consider some point $q \in Q$. Let P' be the subset of P (also of $\mathscr{O}$) such that all points in P' are of distance β' from q. As p is within β of an element of $\mathscr{O}$, it is in some region of the set $R_\mathscr{O}$. Hence, the region that contains p is a partner region for all elements of P'. The statement follows.

Further, as the optimization version of circle covering is known to have no FP-TAS unless $P = NP$ [18], by the nature of the construction in Theorem 3.2, we can be assured of the same result for I-REP-MC.

Corollary 3.1. *I-REP-MC cannot be approximated by a fully polynomial-time approximation scheme (FPTAS) unless $P = NP$.*

Proof. Follows directly from [11] and Theorem 3.2.

So, from the above Theorem and Corollary and Lemma 3.1, we get the following results:

Corollary 3.2. *1. Sub-REP and Sup-REP are NP-Complete.*
2. Sub-REP-MC, Sup-REP-MC, I-REP-MC, Sub-REP-ME, Sup-REP-ME, and I-REP-ME are NP-Hard.
3. Sub-REP-MC, Sup-REP-MC cannot be approximated by a FPTAS unless $P = NP$.

Proof. All follow directly from Lemma 3.1, Theorem 3.2, and Corollary 3.1.

3.4 Algorithms

In this section we devise algorithms to address the optimization problems associated with Sup-REP, Sub-REP, and I-REP. First, we show that these optimization problems reduce to either instances of set-cover (for Sub/Sup-REP-MC) or max-k-cover (for Sub/Sup-REP-ME). These problems are well-studied and there are algorithms that provide exact and approximate solutions. We then provide a new greedy-algorithm for Sub/Sup-REP-MC that also provides an approximation guarantee. This is followed by a discussion of approximation for I-REP-ME for the case where $\alpha = 0$. Finally, we discuss some practical issues dealing with implementation.

3.4.1 Exact and Approximate Solutions by Reduction

In this section we show that the -MC problems introduced earlier in this chapter can be reduced to set-cover and that the -ME problems can reduce to the well-known max-k-cover problem. As these problems have been extensively studied in the core computer science algorithms community, they offer the potential to solve the various region-based geospatial abduction problems introduced earlier in this chapter. Set cover has already been introduced earlier on in Chapter 2. We now present max-k-cover [7], which is often regarded as the dual of set-cover.

Max-k-Cover
INPUT: Set of elements S, family of subsets of S, $\mathscr{H} = H_1, \ldots, H_m$, natural number $k \leq |S|$.
OUTPUT: Subset $\mathscr{H}' \subseteq \mathscr{H}$ s.t. $|\mathscr{H}'| \leq k$ where $|\bigcup_{H_i \in \mathscr{H}'} H_i \cap S|$ is maximized.

The key to showing that Sub/Sup-REP optimization problems can reduce to one of these problems is to determine the family of subsets. We accomplish this as follows: for each region $r \in R$, we find the subset of $\mathscr{O}$ that can be partnered with r. We shall refer to this set as $\mathscr{O}_r$. This gives us the following algorithm for the optimization problems (we simply omit the k parameter for the -MC problems that reduce to Set-Cover):

REDUCE-TO-COVERING($\mathscr{O}$ *set of observations, R set of regions, k natural number*)   returns instance of covering problem $\langle S, \mathscr{H}, k \rangle$

1. For each $r \in R$, find $\mathscr{O}_r$ (*i.e., o is in $\mathscr{O}_r$ iff r sub/super-explains o*)
2. Return $\langle \mathscr{O}, \bigcup_{r \in R} \{\mathscr{O}_r\}, k \rangle$

This algorithm is the analog of the naive KSEP algorithm introduced in Chapter 2. It essentially says that we must perform the following steps.

- For each feasible region $r \in R$, find all the observations "supported" by R (depending on whether we are interested in sub/super-explanations, this means we want to find all observations in R that are within a distance of $[\alpha, \beta]$ of at least one point in R or all points in R). These are the sets $\mathcal{O}_r$ for each $r \in R$.

- We then return the union of all these sets $\mathcal{O}_r$.

It is clear that this algorithm is potentially wasteful, returning regions that can be fed as input to a set covering problem because any set that "covers" all the sets $\mathcal{O}_r$ thus computed yields a potential region that explains the observations. The following result describes the complexity of this algorithm.

Proposition 3.1. *REDUCE-TO-COVERING requires* $O(|\mathcal{O}| \cdot |R|)$ *time.*

Proof. Follows directly from Line 1.

The following theorem shows that REDUCE-TO-COVERING correctly reduces a Sub/Sup-REP optimization problem to set-cover or max-k-cover as appropriate.

Theorem 3.3. *Sub/Sup-REP-MC polynomially reduces to Set-Cover and Sub/Sup-REP-ME polynomially reduces to Max-k-Cover.*

Proof. CLAIM 1: Sub/Sup-REP-MC $\leq_p$ Set-Cover
Consider the instance of set-cover $\langle \mathcal{O}, \bigcup_{r \in R} \{\mathcal{O}_r\} \rangle$ obtained from REDUCE-TO-COVERING($\mathcal{O}, R$).
Let $\mathcal{H}'$ be a solution to this instance of set-cover. ($\Leftarrow$) If R' is a solution to the instance of Sub/Sup-REP-MC, then the set $\bigcup_{r \in R'} \{\mathcal{O}_r\}$ is a solution to set-cover. Obviously, it must cover all elements of $\mathcal{O}$ and a smaller solution to set-cover would indicate a smaller R'—a contradiction. ($\Rightarrow$) Given set $\mathcal{H}'$, let $R'' = \{r \in R | \mathcal{O}_r \in \mathcal{H}'\}$. Obviously, R'' provides a partner for all observations in $\mathcal{O}$. Further, a smaller solution to Sub/Sup-REP-MC would indicate a smaller $\mathcal{H}'$ is possible—also a contradiction.

CLAIM 2: Sub/Sup-REP-ME $\leq_p$ Max-k-Cover
Consider the instance of max-k-cover $\langle \mathcal{O}, \bigcup_{r \in R} \{\mathcal{O}_r\}, k \rangle$ obtained from REDUCE-TO-COVERING($\mathcal{O}, R, k$). Let $\mathcal{H}'$ be a solution to this instance of max-k-cover. ($\Leftarrow$) If R' is a solution to the instance of Sub/Sup-REP-ME, then the set $\bigcup_{r \in R'} \{\mathcal{O}_r\}$ is a solution to max-k-cover. Obviously, both have the same cardinality requirement. Further, if there is a solution to max-k-cover that covers more elements in $\mathcal{O}$, this would imply a set of regions that can be partnered with more observations in $\mathcal{O}$—which would be a contradiction. ($\Rightarrow$) Given set $\mathcal{H}'$, let $R'' = \{r \in R | \mathcal{O}_r \in \mathcal{H}'\}$. Obviously, R'' meets the cardinality requirement of k. Furthermore, a solution to Sub/Sup-REP-ME that allows more observations in $\mathcal{O}$ to be partnered with a region would indicate a more optimal solution to max-k-cover—a contradiction.

This result allows us to leverage any exact approach to the above optimization problems to obtain a solution to an optimization problem associated with Sub/Sup-REP. A straightforward algorithm for any of the optimization problems would run in time exponential in $|\mathcal{O}|$ or k and consider every $|\mathcal{O}|$ or k sized subset of $\bigcup_{r \in R} \{\mathcal{O}_r\}$. Clearly, this is not practical for real-world applications.

Fortunately, there are several well-known approximation techniques for both these problems. First, we address the Sub/Sup-REP-ME problems, both of which reduce to Max-k-Cover. As the Max-k-Cover problem reduces to the maximization of a submodular function over uniform matroid[2], we can leverage the greedy approximation algorithm of [12] to solve our problem. We do so below.

Formally, an arbitrary function $f : X \rightarrow \mathbb{R}$ from some set X to the reals is submodular if and only if for all $X_1, X_2 \subseteq X$, it is the case that if $x \in X - X_2$, then $f(X_1 \cup \{x\}) - f(X_1) \geq f(X_2 \cup \{x\}) - f(X_2)$. Figure 3.4 explains the notion of submodularity; an easy way to explain such functions is given via an intuitive example. Suppose you have a poor man with very few possessions (X_1) and a rich man with many more possessions (X_2). Suppose neither possesses a Ferrari car (x). Giving the poor man the Ferrari would make a greater difference to his net worth (computed via f as a function of the person's possessions) than giving it to the rich man.

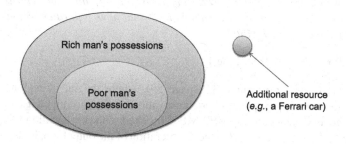

Fig. 3.4 Example of a submodular function. The addition of an expensive vehicle to a rich man's set of possessions would yield a relative increase in net worth far less than the same addition to a poorer man's set of possessions.

GREEDY-REP-ME($\mathscr{O}$ *set of observations*, R *set of regions*, k *natural number*) returns $R' \subseteq R$

1. Let $\mathbf{O} = \bigcup_{r \in R} \{\mathscr{O}_r\}$ (obtained by REDUCE-TO-COVERING)
2. Let $\mathscr{O}' = \mathscr{O}$, set $R' = \emptyset$
3. While $k \neq 0$ loop

    a. Let the element $\mathscr{O}_r$ be the member of $\mathbf{O}$ s.t. $|\mathscr{O}_r \cap \mathscr{O}'|$ is maximized.
    $R' = R' \cup r$
    $\mathscr{O}' = \mathscr{O}' - (\mathscr{O}_r \cap \mathscr{O}')$
    $k - -$

4. Return R'

---

[2] A matroid is a pair (X, I) where X is some set and I is a set of subsets of X (called independent sets) satisfying the following axioms: (i) $\emptyset \in I$, (ii) If $Y \in I$ and $Y' \subseteq Y$, then $Y' \in I$, and (iii) If $Y, Y' \in I$ and $Y' \subset Y$, then there is an element $y \in Y$ such that $(Y' \cup \{y\}) \in I$.

The GREEDY-REP-ME algorithm basically starts by finding all the observations $\mathcal{O}_r$ covered by each region $r \in R$ where R is the set of regions deemed feasible. In order to find a subset of regions of R of cardinality k or less, the algorithm looks at all $\mathcal{O}_r$'s. It initially adds that r into the answer such that $\mathcal{O}_r \cap \mathcal{O}$ is maximized, *i.e.*, in the first iteration of the loop of algorithm GREEDY-REP-ME, it finds an r such that $\mathcal{O}_r$ covers the maximal number of observations in $\mathcal{O}$. In this sense, this algorithm is greedy. This r is added into the solution. As all elements in $\mathcal{O}_r$ have now been "covered" by the insertion of r into the solution, we now only consider elements in $\mathcal{O}' - (\mathcal{O}_r \cap \mathcal{O}')$. The same process is repeated till either $\mathcal{O}'$ is empty or the bound k is reached.

Suppose f denotes the maximum number of observations that can be partnered with a given region. The following result shows an approximation guarantee for our algorithm.

Proposition 3.2. *GREEDY-REP-ME runs in $O(k \cdot |R| \cdot f)$ time and returns a solution such that the number of observations in $\mathcal{O}$ that have a partner region in R' is within a factor $\left(\frac{e}{e-1}\right)$ of optimal.*

Proof. Follows directly from Line 1.

Example 3.6. Consider Example 3.2 (page 59), where the set of regions is $R = \{r_a, r_b, r_c, r_d, r_e, r_f, r_g\}$. Suppose tiger conservationists want to run GREEDY-REP-ME to solve an instance of Sup-REP-ME associated with this situation with $k = 3$. Initially set $\mathcal{O}' = \{o_1, \dots, o_{13}\}$. On the first iteration of the outer loop, it identifies set $\mathcal{O}_{r_c} = \{o_2, o_3, o_4, o_9\}$ where the cardinality of $\mathcal{O}_{r_c} \cap \mathcal{O}$ is maximum. Hence, it picks region r_c. The set $\mathcal{O}' = \{o_1, o_5, \dots, o_8, o_{10}, \dots o_{13}\}$. On the second iteration, it identifies $\mathcal{O}_{r_e} = \{o_5, o_{13}\}$, which intersected with $\mathcal{O}'$ provides a maximum cardinality, causing r_e to be picked. Set $\mathcal{O}'$ is now $\{o_1, o_6, \dots, o_8, o_{10}, \dots, o_{12}\}$. On the last iteration, it identifies $\mathcal{O}_{r_g} = \{o_{11}, o_{12}\}$, again the maximum cardinality when intersected with $\mathcal{O}'$. The element is picked and the solution is r_c, r_e, r_g, and the observations super-explained are $\{o_2, o_3, o_4, o_5, o_9, o_{11}, o_{12}, o_{13}\}$.

Likewise, we can leverage the greedy algorithm for set-cover [26] applied to Sub/Sup-REP-MC. This algorithm is identical to the GREEDY-REP-ME algorithm except in Step (3) where the bound of k is eliminated.

GREEDY-REP-MC($\mathcal{O}$ *set of observations, R set of regions,*) *returns $R' \subseteq R$*

1. Let $\mathbf{O} = \bigcup_{r \in R} \{\mathcal{O}_r\}$ (obtained by REDUCE-TO-COVERING)
2. Let $\mathcal{O}' = \mathcal{O}$, set $R' = \emptyset$
3. While $\mathcal{O}' \neq \emptyset$ loop

    a. Let the element $\mathcal{O}_r$ be the member of $\mathbf{O}$ s.t. $|\mathcal{O}_r \cap \mathcal{O}'|$ is maximized.
       $R' = R' \cup r$
       $\mathcal{O}' = \mathcal{O}' - (\mathcal{O}_r \cap \mathcal{O}')$

4. Return R'

The following result provides approximation guarantees on the solution to the region-based geospatial abduction problem found by the GREEDY-REP-MC algorithm.

Proposition 3.3. *GREEDY-REP-MC runs in $O(|\mathcal{O}| \cdot |R| \cdot f)$ time and returns a solution whose cardinality is within a factor of $1 + \ln(f)$ of optimal.*

Proof. The outer loop of the algorithm iterates no more than $|\mathcal{O}|$ times, while the inner loop iterates no more than $|R|$ times. The time to compare the number of elements in a set $\mathcal{O}_r$ is $O(f)$.

The approximation factor of $1 + \ln(f)$ follows directly from [26]. $\blacksquare$

Example 3.7. Consider the scenario from Example 3.6. To explain all points where a tiger kill has been observed, tiger conservation experts can create an instance of Sup-REP-MC and use GREEDY-REP-MC. The algorithm proceeds just as GREEDY-REP-ME in the first three steps (as in Example 3.6), but will continue on until all observations are super-explained. So, GREEDY-REP-MC proceeds for three more iterations, selecting r_f ($\mathcal{O}_{r_f} = \{o_8, o_{10}\}$), r_d ($\mathcal{O}_{r_d} = \{o_6, o_7\}$), and finally r_a ($\mathcal{O}_{r_a} = \{o_1\}$). The solution returned is:

$$\{r_c, r_e, r_g, r_f, r_d, r_a\}$$

We now focus on speeding up the set-cover reduction via the GREEDY-REP-MC2 algorithm below.

GREEDY-REP-MC2($\mathcal{O}$ *set of observations, R set of regions,*) returns $R' \subseteq R$

1. Let $\mathbf{O} = \bigcup_{r \in R}\{\mathcal{O}_r\}$ (obtained by REDUCE-TO-COVERING)
2. For each observation $o \in \mathcal{O}$, let $GRP_o = \{\mathcal{O}_r \in \mathbf{O} | o \in \mathcal{O}_r\}$
3. For each observation $o \in \mathcal{O}$, let $REL_o = \{o' \in \mathcal{O} | o' \in \bigcup_{\mathcal{O}_r \in GRP_o} \mathcal{O}_r\}$ and let $key_o = |REL_o|$
4. Let $\mathcal{O}' = \mathcal{O}$, set $R' = \emptyset$
5. While $\mathcal{O}' \neq \emptyset$ loop

   a. Let o be the element in $\mathcal{O}$ where key_o is minimal.
   b. Let the element $\mathcal{O}_r$ be the member of GRP_o s.t. $|\mathcal{O}_r \cap \mathcal{O}'|$ is maximized.
   c. If there are more than one set $\mathcal{O}_r$ that meet the criteria of line 5b, pick the set with the greatest cardinality.
   d. $R' = R' \cup r$
   e. For each $o' \in \mathcal{O}_r \cap \mathcal{O}'$, do the following:
      i. $\mathcal{O}' = \mathcal{O}' - o'$
      ii. For each $o'' \in \mathcal{O}' \cap REL_{o'}$, decrement $key_{o''}$

6. Return R'

In the GREEDY-REP-MC2 algorithm, we proceed as follows.

- For any observation $o \in \mathcal{O}$, the set GRP_o is the set of all $\mathcal{O}_r$ where r is a feasible region (*i.e.*, a member of R) that explains o. Thus, $\mathcal{O}_r$ is the set of observations $\mathcal{O}_r$ that contain o and that are explained by some region $r \in R$.

- REL_o is the set of all observations that are contained in sets $\mathcal{O}_r$ that are found in the previous step. Thus, if an observation $o' \in REL_o$, there is at least one region $r \in R$ which explains both o and o'.
- key_o is the size of Rel_o.
- We pick the o such that key_o is minimal, $i.e.$, an o that is "co-explained" as poorly as possible.
- We then find a region r that explains o and that overlaps the set of observations as much as possible. Let $\mathcal{O}_r$ be the set of observations explained by r—if multiple such r's exist, pick the one with the highest cardinality.
- Add r to the "current" answer, and eliminate o from $\mathcal{O}$ as it no longer needs to be explained.
- For every observation $o'' \in \mathcal{O}$ that is explained already by $\mathcal{O}$, we reduce $key_{o''}$ by 1 as one explanation for it has already been found.
- This loop is repeated until all observations are explained.

In the rest of this section, we use Δ to denote the maximum number of different regions that can be partnered with a given observation.

Proposition 3.4. *GREEDY-REP-MC2 runs in* $O(\Delta \cdot f^2 \cdot |\mathcal{O}| + |\mathcal{O}| \cdot \ln(|\mathcal{O}|))$ *time and returns a solution whose cardinality is within a factor of* $1 + \ln(f)$ *of optimal.*

Proof. CLAIM 1: GREEDY-REP-MC2 runs in $O(\Delta \cdot f^2 \cdot |\mathcal{O}| + |\mathcal{O}| \cdot \ln(|\mathcal{O}|))$ time. The pre-processing in lines 1-4 can be accomplished in $O(\Delta + \Delta \cdot f)$ as the size of each GRP_o is bounded by Δ and the size of each REL_o is bounded by $\Delta \cdot f$.

The outer loop of the algorithm iterates $\mathcal{O}$ times. In each loop, the selection of the minimal element (line 5a) can be accomplished in constant time by use of a Fibonacci heap [13] ($i.e.$, storing observations in $\mathcal{O}'$ organized by the value key_o). The next lines of the inner loop (lines 5b-5c) can be accomplished in $O(\Delta)$ time. The next line (line 5d) requires $O(\ln(|\mathcal{O}|))$ time per observation using a Fibonacci heap. However, we can be assured that, during the entire run of the algorithm, this operation is only performed $|\mathcal{O}|$ times (hence, we add an $|\mathcal{O}| \cdot \ln(|\mathcal{O}|)$). The final loop at line 5e occurs after the inner loop and iterates, at most f times. At each iteration, it considers, at most $f \cdot \Delta$ elements. Hence, the overall complexity is:

$$O(|\mathcal{O}| \cdot (\Delta + f^2 \cdot \Delta) + |\mathcal{O}| \cdot \ln(|\mathcal{O}|))$$

The statement of the claim follows.

CLAIM 2: GREEDY-REP-MC2 returns a solution whose cardinality is within a factor of $1 + \ln(f)$ of optimal.
The proof of this claim resembles the approximation proof of the standard greedy algorithm for set-cover (see [5] page 1036).

Let $r_1, \ldots, r_i, \ldots, r_n$ be the elements of R', the solution to GREEDY-REP-MC2, numbered by the order in which they were selected. For each iteration (of the outer loop), let set COV_i be the subset of observations that are partnered for the first time with region r_i. Note that each element of $\mathcal{O}$ is in exactly one COV_i. For each $o_j \in \mathcal{O}$, we define $cost_j$ to be $\frac{1}{|COV_i|}$ where $o_j \in COV_i$. Let R^* be an optimal solution to the instance of Sub/Sup-REP-MC.

CLAIM 2.1: $\sum_{r_i \in R^*} \sum_{o_j \in \mathcal{O}_{r_i}} cost_j \geq |R|$

By the definition of $cost_j$, exactly one unit of cost is assigned every time a region is picked for the solution R. Hence,

$$COST(R) = |R| = \sum_{o_j \in \mathcal{O}} cost_j$$

The statement of the claim follows.

CLAIM 2.2: For some region $r \in R$, $\sum_{o_j \in \mathcal{O}_r} cost_j \leq 1 + \ln(f)$.

Let P be the subset of $\mathcal{O}$ that can be partners with p. At each iteration i of the algorithm, let $uncov_i$ be the number elements in P that do not have a partner. Let $last$ be the smallest number such that $uncov_{last} = 0$. Let $R_P = \{r_i \in R | (i \leq last) \wedge (COV_i \cap P \neq \emptyset)\}$. From here on, we shall renumber each element in R_P as $r_1, \ldots, r_{|R_P|}$ by the order they are picked in the algorithm (*i.e.*, if an element is picked that cannot partner with anything in P, we ignore it and continue numbering with the next available number, we will COV_i and the iterations of the algorithm as well, but do not re-define the set based on the new numbering).

We note that for each iteration i, the number of items in P that are partnered is equal to $uncov_{i-1} - uncov_i$. Hence,

$$\sum_{o_j \in \mathcal{O}_r} cost_j = \sum_{i=1}^{last} \frac{uncov_{i-1} - uncov_i}{|COV_i|}$$

At each iteration of the algorithm, let $PCOV_i$ be the subset of observations that are covered for the first time if region p is picked instead of region r_i. We note, that for all iterations in $1, \ldots, last$, the region p is considered by the algorithm as one of its options for greedy selection. Therefore, as p is not chosen, we know that $|COV_i| \leq |PCOV_i|$. Also, by the definition of $ucov_i$, we know that $|PCOV_i| = ucov_{i-1}$. This gives us:

$$\sum_{o_j \in \mathcal{O}_r} cost_j \leq \sum_{i=1}^{last} \frac{uncov_{i-1} - uncov_i}{ucov_{i-1}}$$

Using the algebraic manipulations of [5] (page 1037), we get the following:

$$\sum_{o_j \in \mathcal{O}_r} cost_j \leq H_{|P|}$$

Where H_j is the jth harmonic number. By definition of the symbol f (maximum number of observations supported by a single partner), we obtain the statement of the claim.

(Proof of Claim 2): Combining claims 1–2, we get $|R| \leq \sum_{r_i \in R^*}(1 + \ln(f))$, which gives us the statement.

While GREEDY-REP-MC2 considers regions in a different order than GREEDY-REP-MC, it maintains the same approximation ratio. This is because the region (in set GRP_o) that is partnered with the greatest number of uncovered observations is selected at each iteration, allowing us to maintain the approximation guarantee. There are two selections at each step: the selection of the observation (in which we use a minimal key value based on related observations) and a greedy selection in the inner loop. Any selection of observations can be used at each step and the approximation guarantee is still maintained. This allows for a variety of different heuristics. Furthermore, the use of a data structure such as a Fibonacci Heap allows us to actually obtain a better time complexity than GREEDY-REP-MC.

Example 3.8. Consider the situation in Example 3.4 where tiger conservation experts are considering regions $R_\mathcal{O} = \{r_1, \dots, r_{33}\}$ that are induced by the set of observations and wish to solve I-REP-MC using GREEDY-REP-MC. On the first iteration of the loop at line 5, the algorithm picks o_8, as $key_{o_8} = 1$. The only possible region to pick is r_{19}, which can only be partnered with o_8. There are no observations related to o_8 other than itself, so it proceeds to the next iteration. It then selects o_6 as $key_{o_6} = 2$ because $REL_{o_6} = \{o_6, o_7\}$. It then greedily picks r_{17} which sub-explains both o_6, o_7. As all observations related to o_6 are now sub-explained, the algorithm proceeds with the next iteration. The observation with the lowest key value is o_5 as $key_{o_5} = 3$ and $REL_{o_5} = \{o_4, o_5, o_{13}\}$. It then greedily picks region r_{21} which sub-explains o_5, o_{13}. The algorithm then reduces the key value associated with o_4 from 4 to 3 and decrements the keys associated with o_{10}, o_{11}, o_{12} (the un-explained observations related to o_{13}) also from 4 to 3. In the next iteration, the algorithm picks o_9 as $key_{o_9} = 3$. It greedily picks r_{12} which sub-explains o_9, o_2. It then decreases key_{o_4} to 2 and also decreases the keys associated with o_1 and o_3. At the next iteration, it picks o_1 as $key_{o_1} = 2$. It greedily selects r_4, which sub-explains o_1, o_3 and decreases the key_{o_4} to 1 which causes o_4 to be selected next, followed by a greedy selection of r_{11}—no keys are updated at this iteration. In the final iteration, it selects o_{10} as $key_{o_{10}} = 3$. It greedily selects r_{25}, which sub-explains all un-explained observations. The algorithm terminates and returns $\{r_{11}, r_{12}, r_{17}, r_{19}, r_{21}, r_{25}\}$.

3.4.2 Approximation for a Special Case

In Section 3.3, we showed that circle covering is polynomially reducible to I-REP-MC. Let us consider a special (but natural) case of I-REP-MC where $\alpha = 0$, *i.e.*, there is no minimum distance between an observation and a partner point that caused it. We shall call this special case I-REP-MCZ. There is a great similarity between this problem and circle-covering. It is trivial to modify our earlier complexity proof to obtain the following result.

Corollary 3.3. *I-REP-MCZ is polynomially reducible to CC.*

Proof. Follows directly from Theorem 3.2.

Furthermore, we can adopt any algorithm that provides a constructive result for circle covering to provide a result for I-REP-MCZ in polynomial time with the following algorithm. Given any point p, it identifies the set $\mathcal{O}_r$ associated with the region that encloses that point.

FIND-REGION($\mathcal{S}$ *space*, $\mathcal{O}$ *observation set*, β *real* , p *point*) returns set $\mathcal{O}_r$

1. Set $\mathcal{O}_r = \emptyset$
2. For each $o \in \mathcal{O}$, if $d(p,o) \leq \beta$ then $\mathcal{O}_r = \mathcal{O}_r \cup \{o\}$
3. Return $\mathcal{O}_r$.

What FIND-REGION does is initially set $\mathcal{O}_r$ to the empty set. It then looks at all observations $o \in \mathcal{O}$. If the observation o is within β units or less from point p, it inserts p into the set $\mathcal{O}_r$.

Proposition 3.5. *The algorithm FIND-REGION runs in $O(|\mathcal{O}|)$ time, and region r (associated with the returned set $\mathcal{O}_r$) contains p.*

Proof. PART 1: FIND-REGION consists of a single loop that iterates $|\mathcal{O}|$ times.

PART 2: Suppose, the region enclosing point p has a different label. Then, there is either a bit in *label* that is incorrectly set to 1 or 0. As only observations which are at a distance of β or less from p have the associated bit position set to 1, then all 1 bits are correct. As we exhaustively consider all observations, the 0 bits are correct. Hence, we have a contradiction.

By pre-processing the regions, we can compute $\mathcal{O}_r$ *a priori* and simply pick a region r associated with the output for FIND-REGION. While there may be more than one such region, any one can be selected as, by definition, they would support the same observations.

Example 3.9. Paleontologists working in a 30×26km area represented by space $\mathcal{S}$ have located scattered fossils of prehistoric vegetation at $\mathcal{O} = \{o_1, o_2, o_3, o_4\}$. Previous experience has led the paleontologists to believe that a fossil site will be within 3km of the scattered fossils. In Figure 3.5, the observations are labeled and circles with radius of 3km are drawn (shown with solid lines). Induced regions $r_1, \ldots, r_6$ are also labeled. As the paleontologists have no additional information, and $\alpha = 0$, they can model their problem as an instance of I-REP-MCZ with $k = 3$. They can solve this problem by reducing it to an instance of circle-covering. The circle-covering algorithm returns three points - p_1, p_2, p_3 (marked with an 'x' in Figure 3.5). Note that each point in the solution to circle-covering falls in exactly one region (when using induced regions). The algorithm FIND-REGION returns the set $\{o_1, o_2\}$ for point p_1, which corresponds with region r_2. It returns set $\{o_3\}$ for p_2, corresponding with r_6 and returns set $\{o_4\}$ for p_3, corresponding with r_5. Hence, the algorithm returns regions r_2, r_6, r_5, which explains all observations.

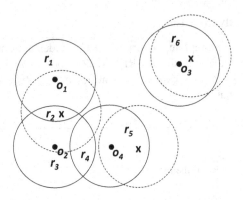

Fig. 3.5 Given the instance of I-REP-MCZ for Example 3.9 as input for circle-covering, a circle-covering algorithm returns points p_1, p_2, p_3 (points are denoted with an "x", dashed circles represent all points within $3km$ from the point).

Any algorithm that provides a constructive result for CC can provide a constructive result for I-REP-MCZ. Because of this one-to-one mapping between the problems, we can also be assured that we maintain an approximation ratio of any approximation technique.

Corollary 3.4. *An a−approximation algorithm for CC is an a-approximation for I-REP-MCZ.*

Proof. Follows directly from Theorem 3.2.

This is useful as we can now use approximation algorithms for CC on I-REP-MCZ. Perhaps the most popular approximation algorithms for CC are based on the "shifting strategy" [18]. To leverage this strategy, we would divide the space, $\mathscr{S}$, into strips of width $2 \cdot \beta$. The algorithm considers groups of ℓ consecutive strips—ℓ is called the "shifting parameter." A local algorithm **A** is applied to each group of strips. The union of all solutions is a feasible solution to the problem. The algorithm then shifts all strips by $2 \cdot \beta$ and repeats the process, saving the feasible solution. This can be done a total of $\ell - 1$ times, and the algorithm simply picks the feasible solution with minimal cardinality. In [18], the following lemma is proved (we state it in terms of I-REP-MCZ—which is done by an application of Corollary 3.4):

Lemma 3.2 (Shifting Lemma [18]). *Let $a_{S(A)}$ be the approximation factor of the shifting strategy applied with local algorithm A and a_A be the approximation factor for the local algorithm. Then:*

$$a_{S(A)} = a_A \cdot \left(1 + \frac{1}{\ell}\right).$$

Furthermore, the shifting strategy can actually be applied twice, solving the local algorithm in squares of size $2 \cdot \beta \cdot \ell \times 2 \cdot \beta \cdot \ell$. This gives the following result:

$$a_{S(S(A))} = a_A \cdot \left(1 + \frac{1}{\ell}\right)^2.$$

A good survey of results based on the shifting strategy can be found in [8], which also provides a linear-time algorithm (this result is later generalized by [9] for multiple dimensions). The following result leverages this for I-REP-MCZ by Corollary 3.4 (and is proved in [9]).

Proposition 3.6. *I-REP-MCZ can be solved with an approximation ratio of $x \cdot \left(1 + \frac{1}{\ell}\right)^2$ in $O(K_{\ell,\rho} \cdot |\mathcal{O}|)$ time. Where p is the maximum number of points in a finite lattice over a square of side length $2 \cdot \beta \cdot \ell$ s.t. each observation in such a square lies directly on a point in the lattice and $x \in \{3, 4, 5, 6\}$ (and is determined by β, see [8] for details) and $K_{\ell,\rho}$ is defined as follows.*

$$K_{\ell,\rho} = \ell^2 \cdot \sum_{i=1}^{\lceil \ell \cdot \sqrt{2} \rceil^2 - 1} \binom{p}{i} \cdot i$$

An alternative to the shifting strategy leverages techniques used for the related problem of geometric dominating set. In [4], the authors present a $1 + \varepsilon$ approximation that runs in $O(|\mathcal{O}|^{O(\frac{1}{\varepsilon^2} \cdot \lg^2(\frac{1}{\varepsilon}))})$ time.

3.4.3 Practical Considerations for Implementation

We now describe some practical implementation issues. Our primary aim is to find a set of regions that resembles the set of induced regions, $R_\mathcal{O}$. There are several reasons for doing this. One reason is to implement a fast heuristic to deal with I-REP optimization problems, specifically when $\alpha \neq 0$. Another, is that such a set of induced regions in the space may be a starting point for creating a set of regions that may include other data, such as that shown in Example 3.5.

As most GIS systems view space as a set of discrete points, we discretized the space using the REGION-GEN algorithm below. The parameter g is the spacing of a square grid that overlays the space.

The result below specifies the running time complexity of the REGION-GEN algorithm.

Proposition 3.7. *REGION-GEN has a time complexity $\Theta(|\mathcal{O}| \cdot \frac{\pi \cdot \beta^2}{g^2})$.*

Proof. For any given observation, the number of points in the grid that can be in a partnered region is $\frac{\pi \cdot \beta^2 - \alpha^2}{g^2}$. Hence, the first loop of the algorithm and the size of L are both bounded by $|\mathcal{O}| \cdot \frac{\pi \cdot \beta^2}{g^2}$. We note that the lookup and insert operations for the

---

REGION-GEN($\mathscr{S}$ *space*, $\mathscr{O}$ *observation set*, α, β, g *reals*) returns set R

1. Overlay a grid of spacing g on space $\mathscr{S}$. With each grid point, p, associate set $\mathscr{O}_p = \emptyset$. This can easily be represented with an array.
2. Initialize list L of pointers to grid-points.
3. For each $o \in \mathscr{O}$, identity all grid points within distance $[\alpha, \beta]$. For each point p meeting this criteria, if $\mathscr{O}_p = \emptyset$, add p to L. Also, set $\mathscr{O}_p = \mathscr{O}_p \cup \{o\}$
4. For some subset $\mathscr{O}' \subset \mathscr{O}$, let $str(\mathscr{O}')$ be a bit string of length $|\mathscr{O}|$ where every position corresponding to an element of $\mathscr{O}'$ is 1 and all other positions are 0.
5. Let T be a hash table of size $\lceil |\mathscr{O}| \cdot \frac{\pi \cdot \beta^2}{g^2} \rceil$ regions indexed by bit-strings of length $|\mathscr{O}|$
6. For each $p \in L$, do the following:

   a. If $T[str(\mathscr{O}_p)] = $ null then initialize this entry to be a rectangle that encloses point p.
   b. Else, expand the region at location $T[str(\mathscr{O}_p)]$ to be the minimum-enclosing rectangle that encloses p and region $T[str(\mathscr{O}_p)]$.

7. Return all entries in T that are not null.

---

hash table T do not affect the average-case complexity. We assume these operations take constant time based on [5], hence the statement follows.

Let us return to our paleontology example from Example 3.9.

Example 3.10. Consider the scenario from Example 3.9. Suppose the paleontologists now want to generate regions using REGION-GEN instead of using induced regions. The algorithm REGION-GEN overlays a grid on the space in consideration. Using an array representing the space, it records the observations that can be explained by each grid point (Figure 3.6, top). As it does this, any grid point that can explain an observation is stored in list L. The algorithm then iterates through list L, creating entries in a hash table for each subset of observations, enclosing all points that explain the same observation with a minimally-enclosing rectangle. Figure 3.6 (bottom) shows the resulting regions $r_1, \ldots, r_6$.

One advantage to using REGION-GEN is that we already have the observations that a region super-explains stored—simply consider the bit-string used to index the region in the hash table. Another thing that can be done, for use in an algorithm such as GREEDY-MC2, where the regions are organized by what observation they support, can also be easily done during the running of this algorithm at an additional cost of f (the number of observations that can be partnered with a given region). This is done by updating an auxiliary data structure, shown at line 6a.

3.5 Experimental Results

We implemented REGION-GEN and GREEDY-MC2 in approximately 3000 lines of Java code and conducted several experiments on a Windows-based computer with an Intel $x86$ processor. Our goal was to show that solving the optimization problem

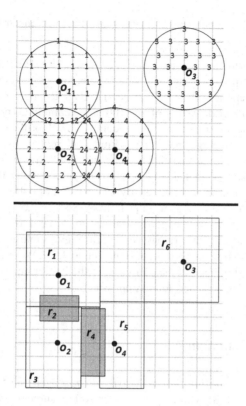

Fig. 3.6 REGION-GEN applied to the paleontology example (Example 3.9). First, it identifies observations associated with grid points (top). It then creates minimally-enclosing rectangles around points that support the same observations (bottom).

Sup-REP-MC would provide useful results in a real-world scenario. We looked at counter-insurgency data from [38] that included data on improvised-explosive device attacks in Baghdad and cache sites where insurgents stored weapons. Under the assumption that the attacks required support of a cache site a certain distance away, could we use attack data to locate cache sites using an instance of Sup-REP-MC solved with GREEDY-MC2 using regions created with REGION-GEN? In our framework, the observations were attacks associated with a cache (which was a partner). The goal was to find relatively small regions that enclosed partners (caches). We evaluated our approach based on the following criteria:

1. Do the algorithms run in a reasonable amount of time?
2. Does GREEDY-MC2 return regions of a relatively small size?
3. Do the regions returned by GREEDY-MC2 usually contain a partner (cache)?
4. Is the partner (cache) density within regions returned by GREEDY-MC2 significantly greater than the partner density of the space?

5. How does the spacing between grid points affect the runtime and accuracy of the algorithms?

Overall, the experiments indicate that REGION-GEN and GREEDY-MC2 satisfactorily meet the requirements above. For example, for our trials considering locating regions with weapons cache sites (partners) in Baghdad given recent IED attacks (observations), with a grid spacing of $g = 100\text{m}$, the combined (mean) run-time on a Windows-based laptop was just over 2 seconds. The algorithm produced (mean) 15.54 regions with an average area of 1.838 km^2. Each region, on average, enclosed 1.739 cache sites. If it did not contain a cache site, it was (on average) 275m away from one. The density of caches within returned regions was 8.09 caches/km²—significantly higher than the overall density for Baghdad of 0.488 caches/km².

The rest of this section is organized as follows. Section 3.5.1 describes the data set we used for our tests and experimental set-up. Issue 1 is addressed in Section 3.5.2. We shall discuss the area (issue 2) of the regions returned in Section 3.5.3 and follow this with a discussion of issue 3 in Section 3.5.4. We shall discuss issue 4 in Section 3.5.5. Throughout all the sections, we shall describe results for a variety of different grid spacings, hence addressing issue 5.

3.5.1 Experimental Setup

We used the *Map of Special Groups Activity in Iraq* available from the Institute for the Study of War [38]. The map plots over 1000 insurgent activities attributed to what are termed as "Special Groups"—groups with access to certain advanced weaponry. This data set—the same one used in Chapter 2—contains events for 21 months between February 2007 and November 2008. The activity types include the following categories.

1. Attacks with probable links to Special Groups
2. Discoveries of caches containing weapons associated with Special Groups
3. Detainments of suspected Special Groups criminals
4. Precision strikes against Special Groups personnel

We use this data for two geographic areas: the Baghdad urban area and the Sadr City district. In our experiment, we will view the attacks by the special groups (item 1) as observations and attempt to determine the minimum set of cache sites (item 3), which we shall view as partners. Hence, a region returned by GREEDY-MC2 encloses a partner iff a cache falls within the region.

For distance constraints, we used a simple algorithm to learn the parameter β (α was set to zero). This was done using the first 7 months of attack data ($\frac{1}{3}$ of the available months) and 14 months of cache data. We used the following simple algorithm, FIND-BETA, to determine these values. Note we set β_{max} to 2.5km.

We ran the experiments on a Lenovo T400 ThinkPad laptop with a 2.53 GHz Intel Core 2 Duo T9400 processor and 4GB of RAM. The computer was running Windows Vista 64-bit Business edition with Service Pack 1 installed.

Algorithm 10 Determines β value from historical data

FIND-BETA($\mathcal{O}_h$ *historical, time-stamped observations,*
$\mathcal{S}_h$ *historical, time-stamped partners,* β_{max} *real*)

1. Set $\beta = \beta_{max}$
2. Set Boolean variable *flag* to TRUE
3. For each $o \in \mathcal{O}_h$, do the following:

    a. For each $p \in \mathcal{S}_h$ that occurs after o, do the following.
        i. Let d be the Euclidean distance function.
        ii. If *flag*, and $d(o,p) \le \beta_{max}$ then set $\beta = d(o,p)$
        iii. If not *flag*, then do the following:
            A. If $d(o,p) > \beta$ and $d(o,p) \le \beta_{max}$ then set $\beta = d(o,p)$

4. Return real β

As the relationship between attacks and cache sites may differ varied on terrain, we ran tests with two different geographic areas. First, we considered the entire Baghdad urban area. Then, we considered just the Sadr City district. We ran FIND-BETA with a β_{max} of 2.5 km on both areas prior to testing the algorithms. There were 73 observations (attacks) for Baghdad and 40 for Sadr City. Table 3.1 shows the exact locations and dimensions of the areas considered.

Area	Lower-Left Latitude	Lower-Left Longitude	E-W Distance	N-S Distance
Baghdad	33.200° N	44.250° E	27 km	25 km
Sadr City	33.345° N	44.423° E	7 km	7 km

Table 3.1 Locations and dimensions of areas considered

We conducted two types of tests: tests focusing on GREEDY-MC2 and tests focusing on REGION-GEN.

For the tests of GREEDY-MC2, we used multiple settings for the grid spacing g. We tested grid spacings at every 10 meter interval in the range of $[70, 1000]$ meters, giving a total of 93 different values for g. Due to the fact that REGION-GEN produces a deterministic result, we ran that algorithm only once per grid setting. However, we ran 100 trials of GREEDY-MC2 for each parameter g. This was done for both Baghdad and Sadr City, giving a total of $18,600$ experiments.

To study the effects of grid-spacing on the run-time of REGION-GEN, we also ran 25 trials for each grid spacing setting for both geographic areas, yielding a total of $4,650$ experiments. To compare the algorithms running with different settings for g in a statistically valid manner, we used ANOVA [14] to determine if the differences among grid spacings were statistically significant. For some test results, we conducted linear regression analysis.

3.5.2 Running Time

Overall, the run times provided by the algorithms were quite reasonable. For example, for the Baghdad trials, 73 attacks were considered for an area of 675km^2. With a grid spacing $g = 100$m, REGION-GEN ran in 2340ms and GREEDY-MC2 took less than 30ms.

For GREEDY-MC2, we found that run-time generally decreased as g increased. For Baghdad, the average run times ranged over $[1.39, 34.47]$ milliseconds. For Sadr City, these times ranged over $[0.15, 4.97]$ milliseconds. ANOVAs for both Baghdad and Sadr City run-times gave p-values of $2.2 \cdot 10^{-16}$, which suggests with well over 99% probability that the algorithm run with different grid settings will result in different run times. We also recorded the number of regions considered in each experiment (resulting from the output of REGION-GEN). Like run-times, we found that the number of regions considered also decreased as the grid spacing increased. For Baghdad, the number of considered regions ranged over $[88, 1011]$. For Sadr City, this number ranged over $[25, 356]$. ANOVAs for both Baghdad and Sadr City number of considered regions gave p-values of $2.2 \cdot 10^{-16}$, which suggests with well over 99% probability that the algorithm run with different grid settings will result in different numbers of considered regions. Note that this is unsurprising as REGION-GEN run deterministically. We noticed that, generally, only grid spacings that were near the same value would lead to the same number of considered regions.

The most striking aspect of the run time/number of regions considered results for GREEDY-MC2 is that these two quantities seem closely related (see Figure 3.7). This most likely results from the fact that the number of regions that can be associated with a given observation (Δ) increases as the number of regions increases. This coincides with our analysis of GREEDY-MC2 (see Proposition 3.4).

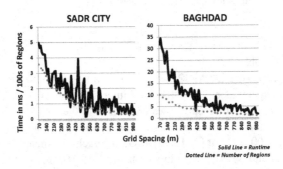

Fig. 3.7 The run time of GREEDY-MC2 in ms compared with the number of regions considered.

We also studied the average run-times for REGION-GEN for various different settings of the grid space g. For Baghdad, the average run times ranged over $[16.84, 9184.72]$ms. For Sadr City, these times ranged over $[0.64, 308.92]$ms. ANOVAs for both Baghdad and Sadr City run-times gave p-values of $2.2 \cdot 10^{-16}$, which suggests with well over 99% probability that the algorithm run with different grid settings will result in different run times. Our analysis of REGION-GEN (See Proposition 3.7) states that the algorithm runs in time $O(\frac{1}{g^2})$. We found striking similarities with this analysis and the experimental results (see Figure 3.8).

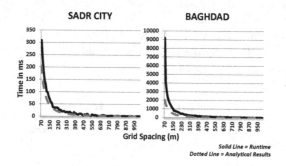

Fig. 3.8 A comparison between analytical $(O(\frac{1}{g^2}))$ and experimental results for the run time of REGION-GEN compared with grid spacing (g).

3.5.3 Area of Returned Regions

In this section, we examine how well the REGION-GEN/GREEDY-MC2 suite of algorithms address the issue of returning regions that are generally small. Although not inherently part of the algorithm, our intuition is that the Sup-REP-MC optimization problem will generally return small regions based on the set R produced by REGION-GEN. The reason for this is that we would expect that smaller regions generally support more observations (note that this is not always true, even for induced regions, but our conjecture is that it is often the case for induced regions or the output of REGION-GEN).

To define "small" we look at the area of a circle of radius β as a basis for comparison. As different grid settings led to different values for β, we looked at the smallest areas. For a given trial, we looked at the average area of the returned regions.

For Baghdad, the average areas ranged over $[0.611, 2.985]$km^2. For Sadr City, these times ranged over $(0.01, 0.576]$km^2. ANOVAs for both Baghdad and Sadr City run-times gave p-values of $2.2 \cdot 10^{-16}$, which suggests with over a 99% probability that the algorithm run with different grid settings will result in different average areas. Plotting the areas compared with the established "minimum area" described earlier in this section clearly shows that REGION-GEN with GREEDY-MC2 produce solutions with an average area that is about half of this value (refer to Figure 3.9).

Overall, there seemed to be little relation between grid spacing and average area
of the returned set of regions—based on grid spacings in $[70, 1000]$m. As an example, we provide screenshots of GREEDY-MC2 for $g = 100$ and $g = 1000$ (Figure 3.10). Anecdotally, we noticed that larger grid spacing led to more "pinpoint"
regions—regions encompassing only one point in the grid (and viewed as having an
area of 0). This is most likely due to the fact that overlaps in the circles around observations points would overlap on fewer grid points for larger values of g. Another
factor is that different settings for g led to some variation of the value β—which
also affects accuracy (note for our analysis we considered only the smallest values
of β as an upper bound for the area (see Figure 3.9).

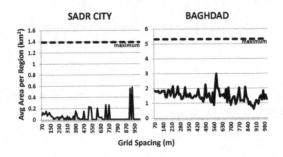

Fig. 3.9 Average areas for solutions provided by REGION-GEN with GREEDY-MC2 for Baghdad and Sadr City.

3.5.4 Regions that Contain Caches

In this section we discuss the issue of ensuring that most of the returned regions
enclose at least one partner (cache in the case of our experiments). One measure
of this aspect is to look at the average number of caches enclosed per region in a
given result. We found that for Baghdad, we generally enclosed more than 1 cache
per region in a given result—this number was in the range $[0.764, 3.25]$. The results
for Sadr City were considerably lower—in the range $[0, 0.322]$. ANOVAs for both
Baghdad and Sadr City gave p-values of $2.2 \cdot 10^{-16}$, which suggests with over a 99%
probability that the algorithm run with different grid settings will result in different
average number of enclosed caches. However, we did not observe an obvious trend
in the data (see Figure 3.11).

As an alternative metric, we look at the number of regions provided by GREEDY-
MC2 that contain at least one partner. Figure 3.13 shows the number of regions returned in the output. For Baghdad, generally fewer than half the regions in the output
will enclose a cache—the number of enclosing regions was in $[1, 8]$, while the total
number of regions was in $[10.49, 22]$. This result, along with the average number

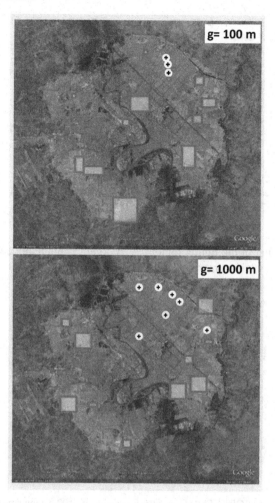

Fig. 3.10 Results from two runs of GREEDY-MC2 - $g = 100m$ (top), $g = 1000m$ (bottom). Pinpoint-regions are denoted with plus-signs. Notice that the average areas of the results are comparable.

of caches enclosed by a region, may indicate that while sometimes GREEDY-MC2 may find regions that enclose many caches, there are often regions that enclose no caches as well. This may indicate that for Baghdad, some attacks-cache relationships conform to our model and others do not. Perhaps there is another discriminating attribute about the attacks not present in the data that may account for this phenomenon. For example, perhaps some attacks were performed by some group that had the capability to store weapons in a cache located further outside the city, or perhaps some groups had the capability to conduct attacks using cache sites that were never found. We illustrate this phenomenon with an example output in Fig-

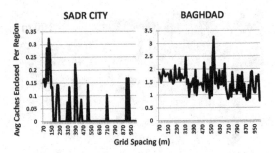

Fig. 3.11 Average caches enclosed per region for Baghdad and Sadr City for various grid-spacing settings.

ure 3.12. Note that in the figure, regions A–E do not contain any cache sites while regions G–I all contain numerous cache sites.

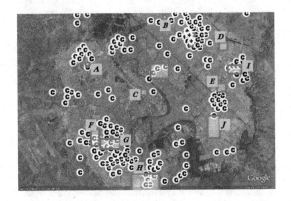

Fig. 3.12 The output of GREEDY-MC2 for Baghdad with $g = 100$m compared with the locations of actual cache sites (denoted with a "C"). Notice that regions A–E do not contain any cache sites while regions G–I all contain numerous cache sites.

For Sadr City, the number of caches that contain one cache was significantly lower—in the range $[0, 2]$—while the total number of returned regions was in $[3, 9.8]$. ANOVAs for both Baghdad and Sadr City gave p-values of $2.2 \cdot 10^{-16}$, which suggests with well over 99% probability that the algorithm, run with different grid settings, will result in different number of regions that enclose a cache location.

We believe that the low numbers for caches enclosed by regions for Sadr City were directly related to the smaller areas of regions. However, the mean of the average area of a returned set of regions was 0 for 49 of the 94 different grid settings (for Sadr City). This means that for the majority of grid settings, the solution consisted only of pinpoint regions (see Section 3.5.3 for a description of pinpoint regions).

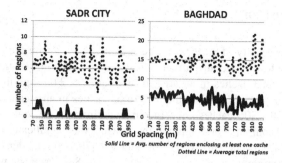

Fig. 3.13 Regions in the output that enclose at least one partner (cache) and total number of regions returned for Baghdad and Sadr City.

Obviously, it is unlikely for a pinpoint region to contain a cache site merely due to its infinitesimally small area. To better account for this issue, we develop another metric: distance to nearest cache. If a region contains a cache, the value for this metric is 0. Otherwise, it is the distance to the closest cache outside of the region. For Baghdad, we obtained distances in $[0.246, 0.712]$km, for Sadr City, $[0.080, 0.712]$km. ANOVAs for both Baghdad and Sadr City gave p-values of $2.2 \cdot 10^{-16}$, which suggests with well over 99% probability that the algorithm run with different grid settings will result in different distances to the nearest cache. Using linear regression, we observed that this distance increases as grid spacing increases. For Baghdad, we obtained $R^2 = 0.2396$ and $R^2 = 0.2688$ for Sadr City. See Figure 3.14 for experimental results and the results of the liner regression analysis.

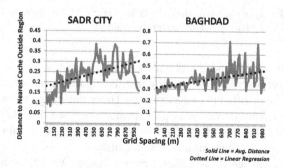

Fig. 3.14 Distance to nearest cache versus grid spacing (in meters).

3.5.5 Partner Density

To consider the density of partners in the regions, we compare the number of enclosed partners to the overall partner density of the area in question. For Baghdad, there were 303 caches in an area measuring 27×24km, giving a density of 0.488 caches/km^2. For Sadr City, there were 64 caches in an area of 7×7km, giving a density of 1.306 caches/km^2. In our experiments, we looked at the cache density for each output. For Baghdad, the density was significantly higher, ranging in $[0.831, 34.9]$ cache/km^2. If we consider $g \in [70, 200]$, the density is between $[7.19, 32.9]$ cache/km^2. For $g = 100$, the density was 8.09 caches/km^2. Most likely due to the issue of pinpoint regions described in Section 3.5.3, the results for Sadr City were often lower than the overall density (in $[0, 31.3]$ cache/km^2). For $g = 100$, the density was 2.08 caches/km^2. We illustrate these results compared with overall cache density in Figure 3.15.

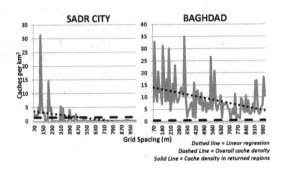

Fig. 3.15 Cache density of outputs produced by GREEDY-MC2 for Baghdad and Sadr City compared with overall cache density and linear-regression analysis.

ANOVAs for both Baghdad and Sadr City gave p-values of $2.2 \cdot 10^{-16}$, which suggests with well over 99% probability that the algorithm run with different grid settings will result in different cache densities. Using linear regression, we observed that this cache density decreases as grid spacing increases. For Baghdad, we obtained $R^2 = 0.1614$ and $R^2 = 0.1395$ for Sadr City. See Figure 3.15 for experimental results and the results of the linear regression analysis.

Although partner density is a useful metric, it does not tell us anything about partners that lie close to a region, although still outside. For example, consider Figure 3.12. Although region A does not enclose any caches, there is a cache just outside. Region B is similar. Also consider the cluster of caches south of region E and north of region J—in this situation it appears as though GREEDY-MC2 mispositioned a region. We include a close-up of region F in Figure 3.16, which encloses a cache, but there are also 4 other caches at a distance of 250m or less.

In order to account for such phenomena, we created an area-quadrupling metric—that is we uniformly double the sides of each region in the output. Then, we cal-

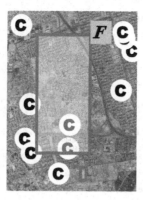

Fig. 3.16 Close-up of region F from Figure 3.12. While region F contains 1 cache, there are 4 other caches < 250m from the boundary of that region. The area-quadrupling metric helps us account for such scenarios.

culated the density of the output with area-quadrupled regions. For Baghdad, this density was in $[0.842, 30.3]$ caches/km^2. For Sadr City, this density was in $[0, 12.3]$ caches/km^2. These results are depicted in Figure 3.17.

As the regions for Sadr City were often smaller than those in Baghdad, we found that the cache density for area-quadrupled regions was often higher for Sadr City (*i.e.*, a region in Sadr City would have nearby cache sites). An example is shown in Figure 3.16.

ANOVAs for both Baghdad and Sadr City gave p-values of $2.2 \cdot 10^{-16}$, which suggests with well over 99% probability that the algorithm run with different grid settings will result in different cache densities for area-quadrupled regions. We also conducted linear regression analysis, and, like the normal partner density, we found that cache density decreases as grid spacing increases. However, this linear analysis was more closely correlated with the data than the analysis for non-area-quadrupled density. For Baghdad, we obtained $R^2 = 0.3171$ (for non-area-quadrupled, we obtained $R^2 = 0.1614$) and $R^2 = 0.3983$ (for non-area-quadrupled, we obtained $R^2 = 0.1395$) for Sadr City. See Figure 3.17 for experimental results and the results of the liner regression analysis.

3.6 Conclusion

In Chapter 2, we developed a formulation of the geospatial abduction problem that assumed that:

- Space was discretized into integer-valued coordinates and handled in a discretized manner as is the case with most real-world GISs; and
- The user desires a set of *points* coming back to him as an explanation.

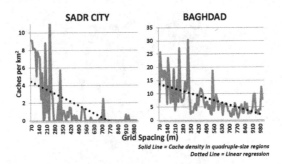

Fig. 3.17 Area quadrupled cache density of output produced by GREEDY-MC2 with linear-regression analysis.

In contrast, in this chapter, we have developed a formalism that:

- Treats space as a continuous two-dimensional set as in this real world (but unlike the way most GIS systems on the market treat space); and
- Returned a set of *regions* to the user.

Returning a set of regions to the user can be highly actionable because each region deemed feasible (and represented as region $r \in R$) may be just large enough so the entity interested in geospatial abduction can act upon the results. For instance, a US military commander looking for weapons caches may know that he cannot search certain regions (for one reason or another). Then he may specify the set R in the input to the region-based geospatial abduction problem to *ignore* such unsearchable regions. In this chapter, R operates much like a feasibility predicate in Chapter 2. However, it goes further in two ways: first, it can be used to specify that *regions* (as opposed to points) can be feasible or infeasible, and second, it can be used to regulate the sizes of the regions that an analyst may want to find. The system would return regions that he can search (*i.e.*, members of R) that offer the best probability of finding a weapons cache. Likewise, in the virus host location detection problem, a public health analyst may set R to consist of only some regions (*e.g.*, a public health expert looking at monkey pox in Rwanda may know that he cannot cross into neighboring countries like Uganda or the Democratic Republic of Congo to eradicate virus hosts). In this case, he may choose only regions $r \in R$ that are within Rwanda and ask the region-based geospatial abduction system to find the best regions in Rwanda for him to target for public health interventions *even though there may be better regions in Uganda or the Democratic Republic of Congo* for him to target with public health interventions.

Thus far, in Chapter 2 and Chapter 3, we have assumed that the adversary is nonchalant and is ignoring our efforts to locate it. This may be reasonable in the case of the virus host detection problem where perhaps mosquitoes and ticks do not have the cognitive capabilities to outwit us. But it is 100% certain that insurgents and terrorists, burglars and other criminals, and even the innocent, but much maligned tiger, have the cognitive capabilities to see what we are doing and adjust their strategy to attempt to outwit us. Tigers are likely to move away from areas of human intervention, just as insurgents in war zones track what we do and react in ways intended to outwit us. This is the focus of the next chapter.

References

1. Alpaydin, E.: 2010. *Introduction to Machine Learning*. MIT Press, 2 edition, 2010.
2. Brantingham, P., Brantingham, P.: 2008. Crime Pattern Theory. In Enviromental Criminology and Crime Analysis, R. Wortley and L. Mazerolle, Eds., pages 78–93.
3. , Bylander, T., Allemang, D., Tanner, M., Josephson, J.R.: 1991. The Computational Complexity of Abduction, Artificial Intelligence.
4. Liao, C., Hu, S.: 2009. Polynomial time approximation schemes for minimum disk cover problems, Journal of Combinatorial Optimization.
5. Cormen, T.H., Leiserson, C.E., Rivest, R.L., Stein, C.: 2001. Introduction to Algorithms. MIT Press, second edition, 2001.
6. Eiter, T., Gottlob, G.: 1995. The complexity of logic-based abduction, J. ACM, 42, 1, pages 3–42.
7. Feige, U.: 1998. A threshold of ln n for approximating set cover, J. ACM, 45, 4, pages 634–652.
8. Franceschetti, M., Cook, M., Bruck, J.: 2004. A Geometric Theorem for Network Design, IEEE Transactions on Computers, 53, 4, pages 483–489.
9. Fu, B, Chen, Z., Abdelguerfi, M.: 2007. An Almost Linear Time 2.8334-Approximation Algorithm for the Disc Covering Problem, AAIM '07: Proceedings of the 3rd international conference on Algorithmic Aspects in Information and Management, pages 317–326, Springer-Verlag.
10. Hochbaum,D.S.,Maass, W.: 1985. Approximation schemes for covering and packing problems in image processing and VLSI, J. ACM, 32, pages 130–136.
11. Megiddo, N., Supowit,K.J.: 1984. On the Complexity of Some Common Geometric Location Problems, SIAM Journal of Computing, 13, 1, pages 182–196.
12. , Nemhauser, G. L., Wolsey, L. A., Fisher, M.L.: 1978. An analysis of approximations for maximizing submodular set functions I, Mathematical Programming, 14, 1, pages 265–294.
13. Fredman, M.L., Tarjan, R.E.: 1987. Fibonacci heaps and their uses in improved network optimization algorithms. Journal of the ACM, 34(3):596–615, July 1987.
14. Freedman, D., Purves, R., Pisani, R.: 2007. Statistics. W.W. Norton and Co., 4 edition.
15. Garey, M.R., Johnson, D.S.: 1979. Computers and Intractability; A Guide to the Theory of NP-Completeness. W. H. Freeman & Co., New York, NY, USA.
16. Hochbaum, D.S.: 1982. Approximation Algorithms for the Set Covering and Vertex Cover Problems. SIAM Journal on Computing, 11(3):555–556.
17. Hochbaum, D.S.: 1997. *Approximation Algorithms for NP-Complete Problems*. PWS Publishing Co., 1997.
18. Hochbaum, D.S., Maass, W.: 1985 Approximation schemes for covering and packing problems in image processing and vlsi. Journal of the ACM, 32:130–136.
19. Jia, L., Rajaraman, R. Suel, T.: 2002. An efficient distributed algorithm for constructing small dominating sets. Distrib. Comput., 15(4):193–205.
20. Johnson, D.S.: 1982. The np-completeness column: An ongoing guide. Journal of Algorithms, 3(2):182–195, 1982.
21. Karp, R.: 1972. Reducibility Among Combinatorial Problems. In R. E. Miller and J. W. Thatcher, editors, Complexity of Computer Computations, page 85-103.
22. Kuhn, F., Wattenhofer, R.: 2003. Constant-time distributed dominating set approximation. In In Proc. of the 22 nd ACM Symposium on the Principles of Distributed Computing (PODC, pages 25–32.

23. Lu, J., Nerode, A., Subrahmanian, V.S.: 1996. Hybrid Knowledge Bases, IEEE Transactions on Knowledge and Data Engineering, 8, 5, pages 773-785.

24. Lund, C., Yannakakis, M.: 1994. On the hardness of approximating minimization problems. Journal of the ACM, 41(5):960–981.

25. Papadimitriou, C.H.: 1981. Worst-Case and Probabilistic Analysis of a Geometric Location Problem, *SIAM J. Comput.*, 10(3):542–557.

26. Paschos, V.T.: 1997. A survey of approximately optimal solutions to some covering and packing problems. ACM Comput. Surv., 29(2):171–209.

27. Reggia, J.A., Peng, Y.: 1990. Abductive inference models for diagnostic problem-solving. Springer-Verlag New York, Inc., New York, NY, USA.

28. Rimoin, A. *et al.*: Endemic Human Monkeypox, Democratic Republic of Congo, 2001-2004, Emerging Infectious Diseases, 13, 6, pages 934–937, 2007.

29. Rossmo, D. K., Rombouts, S.: 2008. Geographic Profiling. In Enviromental Criminology and Crime Analysis, R. Wortley and L. Mazerolle, Eds. pages 136-149.

30. H. Samet.: The Design and Analysis of Spatial Data Structures, Addison Wesley, 1989.

31. Shakarian, P., Subrahmanian, V.S., Sapino, M.L. SCARE: A Case Study with Baghdad, Proc. 2009 Intl. Conf. on Computational Cultural Dynamics (eds. D. Nau, A. Mannes), Dec. 2009, AAAI Press.

32. Shakarian, P., Subrahmanian, V.S., Sapino, M.L. 2012. GAPS: Geospatial Abduction Problems, ACM Transactions on Intelligent Systems and Technology (TIST), 3, 1, to appear.

33. Shakarian, P., Subrahmanian, V.S. Region-based Geospatial Abduction with Counter-IED Applications, accepted for publication in: Wiil, U.K. (ed.).Counterterrorism and Open Source Intelligence, Springer Verlag Lecture Notes on Social Networks, to appear, 2011.

34. Shakarian, P., Nagel, M., Schuetzle, B., Subrahmanian, V.S. 2011. Abductive Inference for Combat: Using SCARE-S2 to Find High-Value Targets in Afghanistan, in Proc. 2011 Intl. Conf. on Innovative Applications of Artificial Intelligence, Aug. 2011, AAAI Press.

35. Shakarian, P., Dickerson, J., Subrahmanian, V.S. 2012. Adversarial Geospatial Abduction Problems, ACM Transactions on Intelligent Systems and Technology (TIST), to appear.

36. Singh, M., Joshi, P.K., Kumar,M., Dash, P.P., Joshi, B.D.: Development of tiger habitat suitability model using geospatial tools: a case study in Achankmar Wildlife Sanctuary (AMWLS), Chhattisgarh India, Env. Monitoring and Assessment journal, Vol. 155, pages 555-567, 2009.

37. US Army: *Intelligence Preparation of the Battlefiled (US Army Field Manual)*, FM 34-130 edition, 1994.

38. "Map of Special Groups Activity in Iraq, Institute for the Study of War", Institute for the Study of War, 2008.

39. Vazirani, V.V.: 2004. Approximation Algorithms. Springer, March 2004.

Chapter 4
Geospatial Abduction with Adaptive Adversaries

Paulo Shakarian, V.S. Subrahmanian, John P. Dickerson

Abstract In this chapter, we focus on the problem of geospatial abduction in the presence of an adversary who understands how we are reasoning about his behavior. For instance, consider an insurgent group carrying out Improvised Explosive Device (IED) attacks on US soldiers. Such an adversary may wish to carry out its attacks and select its cache locations (to support those attacks) in a way that it believes will most likely evade detection. How can an agent (*e.g.*, US forces) anticipate this kind of reasoning by the adversary and find optimal locations to search for weapons caches? In this chapter, we develop a framework to express both the adversary's problem and the agent's problem via the paradigm of Stackelberg games. We formally specify the Optimal Adversary Strategy (OAS) problem, allowing the adversary to find a set of cache locations to minimize (what it believes) to be the probability of being discovered. We describe results on the computational complexity of OAS and algorithms to efficiently compute OAS. As the situation is modeled as a Stackelberg game, the agent (*e.g.*, US forces) takes the final action (*e.g.*, search for the IED caches). The agent can decide where to search *after* considering the space of options that the adversary has and after considering how the adversary might act in order to evade detection. We formalize this as the Maximal Counter-Adversary (MCA) strategy. We describe results on the computational complexity of MCA, as well as algorithms to efficiently compute MCA. These include algorithms that provide guaranteed polynomial approximations to MCA. We describe experimental results about the running time, accuracy, and quality of solutions found by the algorithms to compute OAS and MCA.

4.1 Introduction

We begin by reconsidering several example scenarios given in Chapter 1 where geospatial abduction is required. While all of these scenarios involve an agent (who we implicitly treat as the "good guy") and an adversary (the "bad guy"), we see that some of these scenarios, but not all, represent cases where an adversary can intelligently anticipate what an agent might do, and will take steps to avoid any negative effects of the agent.

- **IED Cache Detection Problem.** Here, US forces (the agent) are trying to find the locations of weapons caches used by insurgents to carry out IED terrorist attacks against civilians and/or US forces. The adversary is constantly adapting its tactics by observing what US soldiers do after any given IED attack. It is therefore clear that after some period of observation of how US forces are searching for IED weapons caches, they will understand at least some elements of the general approach described in Chapters 2 and 3. They may not understand the underlying mathematics, and they will certainly not have the feasibility predicates and lower bound and upper bound cutoff distances used, but they will be able to watch what the US military does on the ground and infer that after attacks, US troops search certain regions and not others. The adversary can therefore be expected to *adapt* his attacks to avoid detection, based on the model of search behavior exhibited by US troops that he may be able to monitor. The problem for the agent (US troops) is to *anticipate* how the adversary might adapt, and use that anticipatory knowledge to discover the location(s) of weapons caches supporting the terror attacks carried out by the adversary.
- **The Tiger Detection Problem.** In the tiger detection problem, we are interested in finding preferred locations where the tiger prefers to reside, based on locations of tiger kills and information about the suitability of various regions on the ground for a tiger dwelling. However, the tiger is a solitary and intelligent animal who would vastly prefer to stay away from human contact. Tigers—and other animals—have, in recent years, been found occasionally in habitats that are different from the ones they usually inhabit. Can wildlife conservationists determine how a tiger is likely to adapt its pattern of behavior as we attempt to search for it, based on the location of tiger kills and habitat information? The ability to do this would significantly enhance tiger conservation efforts.
- **The Criminal Identification Problem.** Burglars, serial killers, and other criminals have a clear interest in avoiding being found by geospatial abduction methods. Should they learn and/or understand what tools and investigative techniques law enforcement officials have at their disposal, then they can adapt their own behavior to minimize the probability of being discovered. Of course, law enforcement officers would like to anticipate how criminals might seek to evade them, and accordingly adjust their own strategy to hunt down the criminals involved.

In contrast to the above problems where the agent seeks to solve a geospatial abduction problem in the presence of an adversary which is adapting its behavior, the Virus Host Identification Problem involves an adversary that is certainly changing—but in its physical makeup rather than in a geospatial sense. Hence, the work described in this chapter does not directly apply to this situation.[1]

---

[1] Some of the results in this chapter may still prove useful in situations like the Virus Host Identification problem. Consider the non-deterministic geospatial abduction algorithms from Chapter 2 such as GREEDY-KSEP-OPT2. Suppose we run one of these algorithms n times, creating n expla-

In the rest of this chapter, we will first describe how geospatial abduction problems can be viewed as two-player games. Then, we will define methods by which an adversary (*e.g.*, insurgents carrying out IED attacks) can best position partner locations (*e.g.*, locations of weapons caches) so as to minimize discovery if the work in the previous sections were to be used. Of course, the agent performing such game theoretic reasoning would like to minimize the probability of being outsmarted by the adversary, so our next section will focus on how to maximize the adversary's probability of being successful. All of these sections will include complexity results, algorithms, and in some cases, approximation algorithms and implementation hints. Finally, we will describe a suite of experiments we have conducted showing that these algorithms work well in a real-life IED cache detection problem using real data drawn from Baghdad.

4.2 Geospatial Abduction as a Two-Player Game

Throughout this chapter, we view geospatial abduction as a two-player game which, like many results in game theory, follows the cyclic outline given below.

1. The *adversary* chooses a set $\mathcal{O}$ of locations where he/it will carry out certain actions that the *agent* can observe.
2. In order to carry out these attacks, the *adversary* tries to find a set of locations that constitute an explanation $\mathcal{E}_1$ of the set $\mathcal{O}$ of observations.
3. The *agent* can detect explanation $\mathcal{E}_1$ using standard geospatial abduction as described in Chapters 2 and 3 (and the *adversary knows this*), so the *adversary* tries to find an explanation $\mathcal{E}_2$ where the *agent* is less likely to detect it.
4. The *agent*, on the other hand, quickly says to himself: "Aha. If the adversary were smart, he would try not to put the explanation at locations in $\mathcal{E}_1$." Instead, putting himself in the adversary's shoes, the *agent* quickly detects that the *adversary* would put the explanation at locations in $\mathcal{E}_2$.
5. The *adversary* can now say "Aha, but if the agent were smart, he would realize that I would not be dumb enough to put the explanation in locations $\mathcal{E}_2$, so he would reason about what I might do and would arrive at the conclusion that I (the *adversary*) would put them at location $\mathcal{E}_3$. At this point, the *agent* would perform a similar analysis (reasoning that the *adversary* would not put it at $\mathcal{E}_2$, etc). This kind of reasoning degenerates into a stage of infinite regress with the *agent* and the *adversary* endlessly trying to stay one step ahead of each other.

Other than purely theoretical interest, the situation where such an infinite regress occurs is neither realistic nor likely. There is so much noise in the real world that

---

nations (some of which may be the same). We can view the virus as having a mixed strategy where it uniformly "selects" one of these n explanations. By framing this as an instance of a Maximal Counter-Adversary Problem (MCA) with an imposed cardinality constraint k (described later in this chapter), an agent can then select the k locations that maximize his payoff with respect to the virus selecting one of the n explanations using a uniform probability distribution.

going too far down this alternating *agent-adversary* reasoning pattern is likely to be extremely complex and not likely to lead to good accuracy or good running time in a real world setting.[2] As a consequence, in this chapter we draw the line at the situation up to point (4) above, *i.e.*, the *adversary* decides to reason up to stage (4) above. But to account for noise, we need to introduce a probabilistic model of how the *adversary* and *agent* reason about each other, together with utilities explaining the value of various situations for them.

4.2.1 Strategies and Rewards

At the end of the day, each player (*agent* or *adversary*) must choose a *strategy* which is merely a subset of the space $\mathscr{S}$.

- When the player considered is the *agent*, the strategy intuitively represents a set of locations that the *agent* believes is the explanation. In the case of the IED detection example, it may be the set of locations that US forces search for an IED weapons cache. In the case of the tiger detection scenario, it may be the set of locations that wildlife experts search for the tiger. The *agent*'s strategy is unknown to the *adversary*.
- When the player considered is the *adversary*, the strategy is the set of locations chosen by the *adversary* to be the true explanation for the observations the agent is causing. For instance, in the tiger detection scenario, the tiger's strategy might be the set of places the tiger dwells before or after making his kills (observations). The *adversary*'s strategy is unknown to the *agent*.

Though "strategy" and "observation" are defined identically, we use separate terms to indicate our intended use. Throughout this chapter, we use $\mathscr{A}$ (resp. $\mathscr{B}$) to denote the strategy of the adversary (resp. agent).

Given a pair $(\mathscr{A}, \mathscr{B})$ of adversary-agent strategies, a reward function measures how similar the two sets are. The more similar the two strategies are, the better it is for the agent. As reward functions can be defined in many ways, we choose an axiomatic approach so that our framework applies to many different reward functions including ones that people may invent in the future.

Definition 4.1 (Reward Function). A *reward function* is any function $\mathbf{rf} : 2^{\mathscr{S}} \times 2^{\mathscr{S}} \to [0, 1]$ that for any k-explanation $\mathscr{A} \neq \emptyset$ and set $\mathscr{B} \subseteq \mathscr{S}$, the function satisfies:

1. If $\mathscr{B} = \mathscr{A}$, then $\mathbf{rf}(\mathscr{A}, \mathscr{B}) = 1$
2. For $\mathscr{B}, \mathscr{B}'$ then
   $$\mathbf{rf}(\mathscr{A}, \mathscr{B} \cup \mathscr{B}') \leq \mathbf{rf}(\mathscr{A}, \mathscr{B}) + \mathbf{rf}(\mathscr{A}, \mathscr{B}') - \mathbf{rf}(\mathscr{A}, \mathscr{B} \cap \mathscr{B}')$$

The basic intuition behind the reward function is that the more the strategy of the agent resembles that of the adversary, the closer the reward is to 1. Axiom 1 says

---

[2] Our complexity results suggest this may be a #P-hard problem

that if the agent's strategy is the same set as the adversary's, then the reward is the maximum possible. *Thus, the magnitude of the reward function always applies to the reward received by the* agent; if the agent guesses precisely where the adversary's chosen explanation is, then the agent gets the maximum possible reward of 1.

Axiom 2 says that adding a point to $\mathcal{B}$ cannot increase the reward to the agent if that point is already in $\mathcal{B}$, *i.e.*, double-counting of rewards is forbidden.

A reader might wonder why certain natural ideas do not count as valid axioms for a reward function. For instance, why is $\mathbf{rf}(\mathcal{A}, \emptyset) = 0$ not an axiom? After all, it could be argued that if the agent does nothing at all, shouldn't there be a zero reward? This is not necessarily true. We can imagine cases where doing something is worse than doing nothing. For instance, in the IED application, the reward to the US military for searching some location (*e.g.*, the house of the Prime Minister of some country) might significantly outweigh the advantage of searching it, especially if there was an error and the hypothetical Prime Minister's house were to be completely empty of any suspicious material. This same argument also explains why reward functions are not necessarily monotonic in the second argument (as the empty strategy for the agent in the preceding discussion could have a higher reward for the agent than the strategy of searching the putative Prime Minister's house).

Nevertheless, there will be cases where some (or many) useful reward functions set $\mathbf{rf}(\mathcal{A}, \emptyset) = 0$ and/or are monotonic in nature. We will consider these later in the chapter. Our next step is to state that rewards associate a simple *payoff* for each player.

Observation 4.2.1 *Given adversary strategy $\mathcal{A}$, agent strategy $\mathcal{B}$, and reward function $\mathbf{rf}$, the payoff for the agent is $\mathbf{rf}(\mathcal{A}, \mathcal{B})$ and the payoff for the adversary is $-\mathbf{rf}(\mathcal{A}, \mathcal{B})$.*

Thus, payoffs are positive for the agent and negative for the adversary. It is therefore easy to see that for any reward function and pair $(\mathcal{A}, \mathcal{B})$, the corresponding game is a *zero-sum game* [1]. Our complexity analysis assumes all reward functions are polynomially computable. All the specific reward functions we propose in this chapter satisfy this condition.

The following important theorem tells us that every reward function is *submodular*, *i.e.*, the marginal benefit of adding additional points to the agent's strategy decreases as the size of the strategy increases. Submodular functions were defined by us in Chapter 3 — we now adapt this notion of submodularity to the case of binary reward functions and show below that reward functions as defined by us are always submodular.

Proposition 4.1 (Submodularity of Reward Functions). *Every reward function is submodular, i.e., if $\mathcal{B} \subseteq \mathcal{B}'$, and point $p \in \mathcal{S}$ such that $p \notin \mathcal{B}$ and $p \notin \mathcal{B}'$, then $\mathbf{rf}(\mathcal{A}, \mathcal{B} \cup \{p\}) - \mathbf{rf}(\mathcal{A}, \mathcal{B}) \geq \mathbf{rf}(\mathcal{A}, \mathcal{B}' \cup \{p\}) - \mathbf{rf}(\mathcal{A}, \mathcal{B}').$*

Proof. Suppose, by way of contradiction, with $\mathcal{B} \subseteq \mathcal{B}'$, and point $p \in \mathcal{S}$ such that $p \notin \mathcal{B}$ and $p \notin \mathcal{B}'$, then

$$\mathbf{rf}(\mathcal{A}, \mathcal{B} \cup \{p\}) - \mathbf{rf}(\mathcal{A}, \mathcal{B}) < \mathbf{rf}(\mathcal{A}, \mathcal{B}' \cup \{p\}) - \mathbf{rf}(\mathcal{A}, \mathcal{B}')$$

We know that $\mathscr{B}' \cup \{p\} = \mathscr{B}' \cup (\mathscr{B} \cup \{p\})$. Hence:

$$\mathbf{rf}(\mathscr{A}, \mathscr{B} \cup \{p\}) - \mathbf{rf}(\mathscr{A}, \mathscr{B}) < \mathbf{rf}(\mathscr{A}, \mathscr{B}' \cup (\mathscr{B} \cup \{p\})) - \mathbf{rf}(\mathscr{A}, \mathscr{B}')$$

Also, we know that $\mathscr{B} = (\mathscr{B} \cup \{p\}) \cap \mathscr{B}'$, so we get:

$$\mathbf{rf}(\mathscr{A}, \mathscr{B} \cup \{p\}) - \mathbf{rf}(\mathscr{A}, (\mathscr{B} \cup \{p\}) \cap \mathscr{B}') < \mathbf{rf}(\mathscr{A}, \mathscr{B}' \cup (\mathscr{B} \cup \{p\})) - \mathbf{rf}(\mathscr{A}, \mathscr{B}')$$

This leads to:

$$\mathbf{rf}(\mathscr{A}, \mathscr{B}') + \mathbf{rf}(\mathscr{A}, \mathscr{B} \cup \{p\}) - \mathbf{rf}(\mathscr{A}, (\mathscr{B} \cup \{p\}) \cap \mathscr{B}') < \mathbf{rf}(\mathscr{A}, \mathscr{B}' \cup (\mathscr{B} \cup \{p\}))$$

which is a clear violation of Axiom 2, hence we have a contradiction.

What this result says is that if the adversary's strategy $\mathscr{A}$ is fixed, then the *marginal benefit* of adding another point to a large agent strategy is not as much as adding the same point to a smaller agent strategy. Simply put, in the case of the IED detection application, if an insurgent group has already placed its weapons at various locations and the agent plans to search either a set $\mathscr{B}$ of points or a superset $\mathscr{B}' \supseteq \mathscr{B}$ of places, and then the agent decides to search one more place p, the increase in overall benefit yielded had they decided to search $\mathscr{B}$ (and p in addition) is greater than the increase in benefit they would get if they searched the superset $\mathscr{B}'$ (and p in addition).

4.2.1.1 Penalizing Reward Function

We explained earlier why $\mathbf{rf}(\mathscr{A}, \emptyset) = 0$ is not an axiom. While this is true of many reward functions, we now give a concrete example of a reward function where we penalize the agent for "bad" strategies because in the real world, executing a bad strategy may have bad consequences. We call this the penalizing reward function.

Definition 4.2 (Penalizing Reward Function). Given a distance *dist*, we define the **penalizing reward function, $\mathbf{prf}^{dist}(\mathscr{A}, \mathscr{B})$**, as follows:

$$\frac{1}{2} + \frac{|\{p \in \mathscr{A} \,|\, \exists p' \in \mathscr{B} \text{ s.t. } d(p,p') \leq dist\}|}{2 \cdot |\mathscr{A}|} - \frac{|\{p \in \mathscr{B} \,|\, \nexists p' \in \mathscr{A} \text{ s.t. } d(p,p') \leq dist\}|}{2 \cdot |\mathscr{S}|}$$

The penalizing reward function intuitively works as follows. It starts at 0.5. It then adds to the reward the ratio of the number of points in the adversary's strategy which are within distance *dist* of some point in the agent's strategy to twice the number of points in the adversary's strategy. Intuitively, this ratio is a measure of the effectiveness of the agent in finding locations in the adversary's explanation. After this ratio is added to the reward, the penalizing reward function *penalizes* the agent for points in the agent's explanation that are not within the given distance *dist* of any point in the adversary's explanation. This intuition is captured by the second ratio in the definition of a penalizing reward function.

Let us consider the IED Cache Detection problem and suppose the agent (US forces) chooses to search locations in $\mathscr{B}$, but the actual places where the adversary has placed his caches are in set $\mathscr{A}$. Intuitively, the **prf**^{dist} function gives the agent a reward for all caches within distance *dist* of a cache location contained in the agent's strategy (representing places the agent plans to search), but it penalizes the agent for all locations searched by the agent that are not within distance *dist* of any actual cache placed by the adversary. Thus, it has the effect of forcing the agent to choose where it searches with great care (*e.g.*, to avoid offending local residents in areas that are subject to the search).

In the same vein, in the tiger identification problem, wildlife conservationists will probably incur a cost for each search they make. Unsuccessful searches may have a cost, both in terms of financial cost, as well as in terms of spooking the tiger and making it harder to find.

The result below states that **prf** satisfies the axioms required for reward functions.

Proposition 4.2. prf^{dist} *is a valid reward function for any dist* ≥ 0.

Proof. In this proof, we define $pt1(\mathscr{A},\mathscr{B}), pt2(\mathscr{A},\mathscr{B})$ as follows:

$$pt1(\mathscr{A},\mathscr{B}) = \frac{|\{p \in \mathscr{A} | \exists p' \in \mathscr{B} \text{ s.t. } d(p,p') \leq dist\}|}{2 \cdot |\mathscr{A}|}$$

$$pt2(\mathscr{A},\mathscr{B}) = \frac{|\{p \in \mathscr{B} | \not\exists p' \in \mathscr{A} \text{ s.t. } d(p,p') \leq dist\}|}{2 \cdot |\mathscr{S}|}$$

Hence, **prf**$^{dist}(\mathscr{A},\mathscr{B}) = 0.5 + pt1(\mathscr{A},\mathscr{B}) - pt2(\mathscr{A},\mathscr{B})$. As we know the maximum value of both $pt1(\mathscr{A},\mathscr{B}), pt2(\mathscr{A},\mathscr{B})$ is 0.5, we know that **prf** is in $[0,1]$. As $pt1(\mathscr{A},\mathscr{A}) = 0.5$ and $pt2(\mathscr{A},\mathscr{A}) = 0$, then Axiom 1 is also satisfied. Consider **crf** (Definition 4.5). Later, in Proposition 4.3, we show that this function is submodular, meeting Axiom 2. By Definitions 4.5, we can easily show that $pt1(\mathscr{A},\mathscr{B}) = 0.5 \cdot \mathbf{crf}^{dist}(\mathscr{A},\mathscr{B})$. As $pt1(\mathscr{A},\mathscr{B})$ is a positive linear combination of submodular functions, it is also submodular. Now consider $pt2(\mathscr{A},\mathscr{B})$. Any element added to any set $\mathscr{B}$ has the same effect—it either lowers the value by $\frac{1}{2 \cdot |\mathscr{S}|}$ or does not affect it— hence it is trivially submodular. Therefore, it follows that **prf** is submodular as it is a positive-linear combination of submodular functions.

The following example revisits the burglary application studied earlier in Chapter 3.

Example 4.1. Consider the two-dimensional space shown in Figure 4.1. Suppose this diagram shows a set of observations (o_i's) depicting locations where burglaries occurred. Furthermore, the police are convinced based on extra-theoretic considerations that the burglaries were carried out by the same burglar (*e.g.*, by examining fingerprints or the burglar's modus operandi). All points in this figure that are not shown in black are assumed to be feasible, and certain locations p_i are marked with numbers (just the i for readability). Suppose the burglar's actual places of residence (*e.g.*, home and office) are given by $\mathscr{A} = \{p_{40}, p_{46}\}$ while the set $\mathscr{B} = \{p_{38}, p_{41}, p_{44}, p_{56}\}$ represents locations that that the police wish to search. Suppose we consider distance $dist = 100$ meters. There is only one point in $\mathscr{A}$ that is within 100 meters of a point in $\mathscr{B}$ (point p_{40}) and 3 points in $\mathscr{B}$ more than 100 meters from any point in $\mathscr{A}$ (points p_{38}, p_{44}, p_{56}). These relationships are shown visually in Figure 4.1. Hence, $\mathbf{prf}^{dist}(\mathscr{A}, \mathscr{B}) = 0.5 + 0.25 - 0.011 = 0.739$.

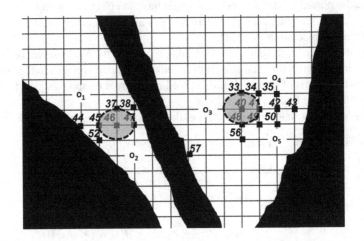

Fig. 4.1 Dashed circles encompass all feasible points within 100 meters from explanation $\{p_{40}, p_{45}\}$. Regions shown in black are deemed infeasible by the supplied feasibility predicate.

Definition 4.3. A reward function is said to be *zero-starting* if $\mathbf{rf}(\mathscr{A}, \emptyset) = 0$.

Thus, a reward function is zero-starting if the agent gets no reward for having an empty strategy. As mentioned earlier in this chapter, not all reward functions should be required to be zero-starting; however, there may be plenty of zero-starting reward functions that are useful in many cases. We now define *monotonic* reward functions.

Definition 4.4. A reward function, **rf**, is **monotonic** if (i) it is zero-starting and (ii) if $\mathscr{B} \subseteq \mathscr{B}'$ then $\mathbf{rf}(\mathscr{A}, \mathscr{B}) \leq \mathbf{rf}(\mathscr{A}, \mathscr{B}')$.

Note that in standard mathematics, monotonic functions in general are not required to satisfy the first condition in the above definition. However, our definition of monotonic reward functions requires them to be zero-starting as well. We now define several example monotonic reward functions, starting with the "cutoff" reward function.

4.2.1.2 Cutoff Reward Function

The intuition behind the *cutoff reward function* **crf** is simple. Suppose we are given a distance *dist* (the "cutoff" distance). The cutoff function looks at the percentage of locations in the adversary's strategy that are within *dist* units of some point in the agent's strategy.

Definition 4.5 (Cutoff Reward Function). Reward function based on a cutoff distance *dist*.

$$\mathbf{crf}^{dist}(\mathscr{A}, \mathscr{B}) := \frac{card(\{p \in \mathscr{A} \,|\, \exists p' \in \mathscr{B} \text{ s.t. } d(p, p') \le dist\})}{card(\mathscr{A})}$$

Intuitively, the cutoff reward function assumes that if the agent's strategy includes a point p' and there is a location p in the adversary's strategy that is within *dist* units of p', then the agent will discover it. Returning to our IED Cache Detection problem, intuitively this says that the agent will find all caches used by insurgents if those caches are within *dist* units of a location that the agent has decided to search. In the case of the Tiger Identification problem, likewise, this says that if wildlife experts decide to search a point p' that is within *dist* units of an actual tiger dwelling, then the wildlife experts will in fact discover this.

Thus, in the case of the IED Cache Detection application, the cutoff reward function intuitively specifies the percentage of enemy caches actually discovered by the agent's strategy, while in the case of the Tiger Identification problem, it specifies the percentage of tiger dwellings that actually exist. The following proposition shows that the cutoff reward function is a valid, monotonic reward function.

Proposition 4.3. $\mathbf{crf}^{dist}$ *is a valid, monotonic reward function for any dist* ≥ 0.

Proof. CLAIM 1: **crf** satisfies reward Axiom 1.
Clearly, if $\mathscr{B} = \mathscr{A}$, then the numerator is $|\mathscr{A}|$, which equals the denominator.

CLAIM 2: **crf** satisfies reward function Axiom 2.
Suppose, by way of contradiction, there exists explanations $\mathscr{B}, \mathscr{B}'$ such that $\mathscr{B} \cup \mathscr{B}'$ is an explanation and $\mathbf{crf}^{dist}(\mathscr{A}, \mathscr{B} \cup \mathscr{B}') > \mathbf{crf}^{dist}(\mathscr{A}, \mathscr{B}) + \mathbf{rf}(\mathscr{A}, \mathscr{B}') - \mathbf{rf}(\mathscr{A}, \mathscr{B} \cap \mathscr{B}')$. Therefore, $card(\{p \in \mathscr{A} \,|\, \exists p' \in \mathscr{B} \cup \mathscr{B}' \text{ s.t. } d(p, p') \le dist\})$ is greater than $card(\{p \in \mathscr{A} \,|\, \exists p' \in \mathscr{B} \text{ s.t. } d(p, p') \le dist\}) + card(\{p \in \mathscr{A} \,|\, \exists p' \in \mathscr{B}' \text{ s.t. } d(p, p') \le dist\}) - card(\{p \in \mathscr{A} \,|\, \exists p' \in \mathscr{B} \cap \mathscr{B}' \text{ s.t. } d(p, p') \le dist\})$. We have a contradiction; indeed, by basic set theory we see that both sides of this strict inequality are actually equal.

CLAIM 3: **crf** is zero-starting.
Clearly, if $\mathscr{B} = \emptyset$, the numerator must be 0, and the statement follows.

CLAIM 4: **crf** is monotonic.
Suppose, by way of contradiction, there exists $\mathscr{B} \subseteq \mathscr{B}'$ such that $\mathbf{rf}(\mathscr{A}, \mathscr{B}) > \mathbf{rf}(\mathscr{A}, \mathscr{B}')$. Then $card(\{p \in \mathscr{A} \,|\, \exists p' \in \mathscr{B} \text{ s.t. } d(p, p') \le dist\}) > card(\{p \in \mathscr{A} \,|\, \exists p' \in \mathscr{B}' \text{ s.t. } d(p, p') \le dist\})$. Clearly, this is not possible as $\mathscr{B} \subseteq \mathscr{B}'$ and we have a contradiction.

The following example illustrates how the **crf** reward function works.

Example 4.2. Consider Example 4.1. Here, $\mathbf{crf}^{dist}(\mathscr{A}, \mathscr{B})$ returns 0.5 as one element of $\mathscr{A}$ is within 100 meters of an element in $\mathscr{B}$.

4.2.1.3 Falloff Reward Function

The cutoff reward function uses *dist* to decide whether a location in the adversary's strategy will be discovered by the agent or not. Thus, even if an adversary's chosen location is just an inch outside the *dist* bound from the closest point in the agent's strategy, this function would say that the adversary's location would not be found. This may not always be realistic.

In contrast, the falloff reward function **frf** defined below says that for each location *p* in the adversary's strategy (*e.g.*, where the insurgent group chooses to place its IED weapons locations, or where the tiger actually resides in practice), the probability that the agent will discover this location is inversely proportional to the distance of *p* from the nearest point p' that is in the agent's strategy.

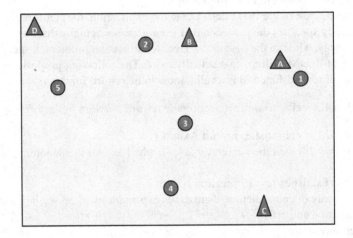

Fig. 4.2 Example adversary and agent strategies for the falloff reward function (**frf**). The agent's strategy consists of points marked by blue circles, while the adversary's strategy consists of the red triangles.

Figure 4.2 shows a simple example. In this figure, we look at each location in the adversary's strategy; in the IED detection application, for example, these are the locations where the insurgents decided to place their weapons caches. For each such adversary location, we find the location in the agent's strategy that is closest to it. The table below summarizes this situation.

	Adversary Location	Nearest Agent Location	Distance
	A	1	1
Example 4.3.	B	2	2
	C	4	5
	D	5	4

The falloff reward function assigns a reward to the agent that is inversely proportional to the distances between each *A* and 1, *B* and 2, *C* and 4, and *D* and 5.

As long as the reward increases as the distances in the above table decreases, we have a function that rewards the agent for a strategy which comes "close" (in terms of distance) to the strategy of the adversary. The falloff reward function defined below implements this intuition in one way—many other ways to achieve the same intuition, albeit with slightly different definitions of the falloff reward function, are possible.

Definition 4.6 (Falloff Reward Function). Reward function with value based on minimal distances between points.

$$\mathbf{frf}(\mathscr{A},\mathscr{B}) := \begin{cases} 0 & \text{if } \mathscr{B} = \emptyset \\ \sum_{p \in \mathscr{A}} \frac{1}{|\mathscr{A}| + \min_{p' \in \mathscr{B}}(d(p,p')^2)} & \text{otherwise} \end{cases}$$

with $d(p,p') := \sqrt{(p_x - p'_x)^2 + (p_y - p'_y)^2}$. In this case, the agent's reward is inversely proportional to the square of the distance between points, as the search area required grows proportionally to the square of this distance.

The example below extends the preceding example.

Example 4.4. Let us continue with the situation in the table shown in Example 4.3. In this case, we see that when the agent strategy is $\{1,2,3,4,5\}$ and the adversary's strategy is $\{A,B,C,D\}$. In this case, $|\mathscr{A}| = 4$ as the adversary has placed four caches. Then the falloff reward function returns the following:

$$\frac{1}{4+1^2} + \frac{1}{4+2^2} + \frac{1}{4+5^2} + \frac{1}{4+4^2}$$

which turns out to be $0.2 + 0.125 + 0.034 + 0.05 = 0.409$ which is the value returned by the falloff reward function.

The following result specifies that the falloff reward function **frf** satisfies the axioms to be a reward function—and, additionally, is monotonic.

Proposition 4.4. frf *is a valid, monotonic reward function.*

Proof. CLAIM 1: **frf** satisfies all reward function axioms (*i.e.*, is valid).

Bounds   We must show $\mathbf{rf}(\mathscr{A},\mathscr{B}) \in [0,1]$. For each point $p \in \mathscr{A}$, let $l_p^{\mathscr{B}} = \min_{p' \in \mathscr{B}} d(p,p')^2$. By the definition of the distance function d, we know $0 \leq l_p^{\mathscr{B}} < \infty$. Now let function $f(l_p^{\mathscr{B}}) = \frac{1}{|\mathscr{A}| + \min_{p' \in \mathscr{B}} d(p,p')^2} = \frac{1}{|\mathscr{A}| + l_p^{\mathscr{B}}}$. Since $0 \leq l_p^{\mathscr{B}} < \infty$, we see $0 < f(l_p^{\mathscr{B}}) \leq \frac{1}{|\mathscr{A}|}$. Clearly, the summation over $|\mathscr{A}|$ points $p \in \mathscr{A}$ yields an answer in $\left(0, |\mathscr{A}| \cdot \frac{1}{|\mathscr{A}|}\right] = (0,1] \subset [0,1]$.

Axiom 1   If $\mathscr{B} = \mathscr{A}$, for each $p \in \mathscr{A}$, there exists $p' \in \mathscr{B}$ such that $d(p,p') = 0$. Hence, by the definition of **frf**, $\mathbf{frf}(\mathscr{A},\mathscr{B}) = 1$ in this case.

Axiom 2   We must show that our version of the triangle inequality holds, that is $\mathbf{rf}(\mathscr{A}, \mathscr{B} \cup \mathscr{B}') \leq \mathbf{rf}(\mathscr{A}, \mathscr{B}) + \mathbf{rf}(\mathscr{A}, \mathscr{B}') - \mathbf{rf}(\mathscr{A}, \mathscr{B} \cap \mathscr{B}')$. From above, $\mathbf{rf}(\mathscr{A}, \mathscr{B} \cup \mathscr{B}') = \sum_{p \in \mathscr{A}} f(l_p^{\mathscr{B} \cup \mathscr{B}'})$. For each point $p \in \mathscr{A}$, let $p_* = argmin_{p' \in \mathscr{B} \cup \mathscr{B}'} d(p, p')^2$. Without loss of generality, assume $p_* \in \mathscr{B}$, then $l_p^{\mathscr{B}} = l_p^{\mathscr{B} \cup \mathscr{B}'}$ thus $f(l_p^{\mathscr{B}}) = f(l_p^{\mathscr{B} \cup \mathscr{B}'})$. Since $p_* \in \mathscr{B}$, we have $p_* \in \mathscr{B} \cap \mathscr{B}'$ or $p_* \in \mathscr{B} \cap \bar{\mathscr{B}}'$.

If $p_* \in \mathscr{B} \cap \mathscr{B}'$:   Then $f(l_p^{\mathscr{B} \cap \mathscr{B}'}) = f(l_p^{\mathscr{B}})$. However, since $p_* \in \mathscr{B}'$ we have, as above, $f(l_p^{\mathscr{B}'}) = f(l_p^{\mathscr{B}}) = f(l_p^{\mathscr{B} \cup \mathscr{B}'})$. Thus

$$\sum_{p \in \mathscr{A}} \left[f(l_p^{\mathscr{B}}) + f(l_p^{\mathscr{B}'}) - f(l_p^{\mathscr{B} \cap \mathscr{B}'}) \right] \tag{4.1}$$

$$= \sum_{p \in \mathscr{A}} \left[f(l_p^{\mathscr{B} \cup \mathscr{B}'}) + f(l_p^{\mathscr{B} \cup \mathscr{B}'}) - f(l_p^{\mathscr{B} \cup \mathscr{B}'}) \right] \tag{4.2}$$

$$= \sum_{p \in \mathscr{A}} f(l_p^{\mathscr{B} \cup \mathscr{B}'}) \tag{4.3}$$

So $\mathbf{rf}(\mathscr{A}, \mathscr{B} \cup \mathscr{B}') = \mathbf{rf}(\mathscr{A}, \mathscr{B}) + \mathbf{rf}(\mathscr{A}, \mathscr{B}') - \mathbf{rf}(\mathscr{A}, \mathscr{B} \cap \mathscr{B}')$ for this case.

If $p_* \in \mathscr{B} \cap \bar{\mathscr{B}}'$:   From above, we are still guaranteed that $f(l_p^{\mathscr{B}}) = f(l_p^{\mathscr{B} \cup \mathscr{B}'})$, thus $\mathbf{rf}(\mathscr{A}, \mathscr{B} \cup \mathscr{B}') = \mathbf{rf}(\mathscr{A}, \mathscr{B})$. This reduces our problem to showing $\mathbf{rf}(\mathscr{A}, \mathscr{B}') - \mathbf{rf}(\mathscr{A}, \mathscr{B} \cap \mathscr{B}') \geq 0$. However, $\mathbf{rf}$ is monotonic (shown below); since $\mathscr{B} \cap \mathscr{B}' \subseteq \mathscr{B}'$, then $\mathbf{rf}(\mathscr{A}, \mathscr{B} \cap \mathscr{B}') \leq \mathbf{rf}(\mathscr{A}, \mathscr{B}')$ and our claim holds.

A similar proof holds for the case $p_* \in \mathscr{B}'$.

CLAIM 2: $\mathbf{frf}$ is monotonic and zero-starting. The property of zero-starting follows directly from the definition of $\mathbf{frf}$.

By way of contradiction, assume there is some $\mathscr{B} \subset \mathscr{B}'$ such that $\mathbf{rf}(\mathscr{A}, \mathscr{B}) > \mathbf{rf}(\mathscr{A}, \mathscr{B}')$. Then, as above, $\sum_{p \in \mathscr{A}} f(l_p^{\mathscr{B}}) > \sum_{p \in \mathscr{A}} f(l_p^{\mathscr{B}'})$. However, since $\mathscr{B} \subset \mathscr{B}'$, we have $l_p^{\mathscr{B}} \geq l_p^{\mathscr{B}'}$ for each $p \in \mathscr{A}$. Similarly, $f(l_p^{\mathscr{B}}) \leq f(l_p^{\mathscr{B}'})$ and thus $\sum_{p \in \mathscr{A}} f(l_p^{\mathscr{B}}) \leq \sum_{p \in \mathscr{A}} f(l_p^{\mathscr{B}'})$, which is our contradiction.

4.2.1.4 Weighted Reward Function

In all the specific examples of reward functions presented thus far, all locations in both the agent's strategy and the adversary's strategy are considered to be equally important (though there is a hint that this may not be the case when we discussed searching a Prime Minister's house in the discussion on penalizing reward functions). We now define *weighted reward functions*, where each location p' in the agent's strategy has an associated weight.

Returning to the IED detection example, the weight of searching the Prime Minister's house might be very low, while a national security analyst may set the weight of searching a mosque or the grounds of an extremist *madrasah* (religious school) to be much higher. Likewise, in the case of the tiger detection example, the wildlife conservation expert might set the weight of searching a particular location in a way that is consistent with the suitability of the habitat (*e.g.*, density of ground cover or forest, abundance of prey) for the tiger to inhabit. Thus, the weighted reward function **wrf** assigns a greater reward for being "closer" to points in $\mathscr{A}$ that have high weight than those with lower weights.

Definition 4.7 (Weighted Reward Function). Given weight function $W : \mathscr{S} \to \mathbb{R}^+$, and a cutoff distance *dist* we define the *weighted reward function* to be:

$$\mathbf{wrf}^{(W,dist)}(\mathscr{A},\mathscr{B}) := \frac{\sum_{\{p \in \mathscr{A} \mid \exists p' \in \mathscr{B} \ s.t. \ d(p,p') \leq dist\}} W(p)}{\sum_{p' \in \mathscr{A}} W(p')}$$

Thus, the weighted reward function proceeds as follows. For a given cutoff distance *dist*, it considers each location p in the adversary's strategy and checks if there exists a location p' in the agent's strategy. If so, it adds the weight of p (which intuitively indicates the importance of the adversary's location p from the point of view of the GAP application) to a running total. Once all such locations $p \in \mathscr{A}$ have been considered, it divides the total obtained by the sum of weights of all points in $\mathscr{A}$ to obtain a fractional value of the locations that the agent is expected to discover, should the agent and the adversary use strategies $\mathscr{B}$ and $\mathscr{A}$, respectively.

The following result establishes that the weighted reward function satisfies the axioms required to be a reward function—and, moreover, is monotonic.

Proposition 4.5. $\mathbf{wrf}^{(W,dist)}(\mathscr{A},\mathscr{B})$ *is a valid, monotonic reward function.*

Proof. CLAIM 1: **wrf** satisfies all reward function axioms (*i.e.*, is valid).

Domain   We must show $\mathbf{wrf}^{(W,dist)}(\mathscr{A},\mathscr{B}) \in [0,1]$. As $(\mathscr{B} \cap \mathscr{A}) \subseteq \mathscr{A}$ and W only returns positive values, this function can only return values in $[0,1]$.

Axiom 1   If $\mathscr{B} = \mathscr{A}$, then for each $p \in \mathscr{A}$, there exists $p' \in \mathscr{B}$ such that $d(p,p') = 0$. This causes the numerator to equal $\sum_{p \in \mathscr{B}} W(p)$. As $\mathscr{B} = \mathscr{A}$, the numerator is equivalent to the denominator, so $\mathbf{wrf}(\mathscr{A},\mathscr{B}) = 1$ in this case.

Axiom 2   We must show the inequality $\mathbf{wrf}^{(W,dist)}(\mathscr{A},\mathscr{B} \cup \mathscr{B}') \leq \mathbf{wrf}^{(W,dist)}(\mathscr{A},\mathscr{B}) + \mathbf{wrf}^{(W,dist)}(\mathscr{A},\mathscr{B}') - \mathbf{wrf}^{(W,dist)}(\mathscr{A},\mathscr{B} \cap \mathscr{B}')$. This proof is similar to the proof of Axiom 2 in Proposition 4.3.

CLAIM 2: **wrf** is monotonic and zero-starting.
The property of zero-starting if shown by when $\mathscr{B} = \emptyset$, the numerator must be 0, hence, $\mathbf{wrf}(\mathscr{A},\emptyset) = 0$. By way of contradiction, assume there is some $\mathscr{B} \subset \mathscr{B}'$ such that $\mathbf{wrf}^{(W,dist)}(\mathscr{A},\mathscr{B}) > \mathbf{wrf}^{(W,dist)}(\mathscr{A},\mathscr{B}')$. Then

$$\frac{\sum_{\{p \in \mathscr{A} \mid \exists p' \in \mathscr{B} \ s.t. \ d(p,p') \leq dist\}} W(p)}{\sum_{p' \in \mathscr{A}} W(p')} > \frac{\sum_{\{p \in \mathscr{A} \mid \exists p' \in \mathscr{B}' \ s.t. \ d(p,p') \leq dist\}} W(p)}{\sum_{p' \in \mathscr{A}} W(p')}$$

Since $\mathscr{B} \subset \mathscr{B}'$, we have

$$\frac{\sum_{\{p \in \mathscr{A} \mid \exists p' \in \mathscr{B} \ s.t. \ d(p,p') \leq dist\}} W(p)}{\sum_{p' \in \mathscr{A}} W(p')} >$$

$$\frac{\sum_{\{p\in\mathscr{A}\mid \exists p'\in\mathscr{B}\ s.t.\ d(p,p')\leq dist\}} W(p)}{\sum_{p'\in\mathscr{A}} W(p')} + \frac{\sum_{\{p\in\mathscr{A}'\mid \exists p'\in(\mathscr{B}'\cap\mathscr{B})\ s.t.\ d(p,p')\leq dist\}} W(p)}{\sum_{p'\in\mathscr{A}} W(p')}$$

Where $\mathscr{A}' = \{p\in\mathscr{A}\mid \not\exists p'\in\mathscr{B}\ s.t.\ d(p,p')\leq dist\}$. Hence,

$$0 > \mathbf{wrf}^{(W,dist)}(\mathscr{A}',\mathscr{B}'\cap\mathscr{B})$$

Which violates the first axiom, which was shown to apply to $\mathbf{wrf}^{(W,dist)}$ by Claim 1—a contradiction.

It is easy to see that the weighted reward function is a generalization of the cutoff reward function where all weights are 1.

4.2.2 Incorporating Mixed Strategies

Of course, neither the adversary nor the agent wants to be entirely predictable, as predictability (in the case of the adversary) would mean that the agent has an easy way to uncover its hidden locations, while predictability (in the case of the agent) means the adversary can easily avoid being uncovered. To achieve unpredictability, they are not likely to pick one strategy and stick with it; rather, they are likely to pick strategies in accordance with some probability distribution. In this section, we introduce probability density functions (pdfs) over strategies (or *mixed strategies* as they are commonly referred to in game theory [1]) and introduce the notion of expected reward. We first present *explanation/strategy functions* which return an explanation (resp. strategy) of a certain size for a given set of observations.

Definition 4.8 (Explanation/Strategy Function). An *explanation (resp. strategy) function* is any function $\mathsf{ex\_fcn} : 2^{\mathscr{S}} \times \mathbb{N} \to 2^{\mathscr{S}}$ (resp. $\mathsf{sf} : 2^{\mathscr{S}} \times \mathbb{N} \to 2^{\mathscr{S}}$) that, given a set $\mathscr{O} \subseteq \mathscr{S}$ and $k\in\mathbb{N}$, returns a set $\mathscr{E} \subseteq \mathscr{S}$ such that $\mathscr{E}$ is a k-sized explanation of $\mathscr{O}$ (resp. $\mathscr{E}$ is a k-sized subset of $\mathscr{S}$). Let **EF** be the set of all explanation functions.

Intuitively, all that an explanation function does is to return an explanation of size k, given a set of observations and an integer k as input. In contrast, a strategy function just returns a strategy of size k.

Example 4.5. Continuing with Example 4.1, we now define two functions $\mathsf{ex\_fcn}_1$ and $\mathsf{ex\_fcn}_2$. Given the set $\mathscr{O}$ (defined in Example 4.1) as input and $k \leq 3$, these functions give the following results:

$$\mathsf{ex\_fcn}_1(\mathscr{O},3) = \{p_{42}, p_{45}, p_{48}\}$$
$$\mathsf{ex\_fcn}_2(\mathscr{O},3) = \{p_{40}, p_{46}\}$$

These sets may correspond to explanations from various sources. Perhaps they correspond to the answer of an algorithm that police officials use to solve GAPs. Conversely, they could also be places the burglar thinks would make most sense for him to inhabit.

In theory, the set of all explanation functions can be infinitely large; however, it makes no sense to look for explanations containing more points than $\mathscr{S}$ and so we assume explanation functions are only invoked with $k \leq (M+1) \times (N+1)$.

A strategy function is appropriate for an agent who wants to select points resembling what the adversary selected, but is not required to produce an explanation. Our results typically do not depend on whether an explanation or strategy function is used (when they do, we point it out). Therefore, for simplicity, we use "explanation function" throughout the chapter. In our complexity results, we assume that explanation/strategy functions are computable in constant time.

Both the agent and the adversary do not know the explanation function (*e.g.*, answers to the questions "where is the adversary going to put his weapons caches?" and "where will US forces search for them?") in advance. Thus, they use a pdf over explanation functions to estimate their opponent's behavior, yielding a *mixed strategy*.

Definition 4.9 (Explanation Function Distribution). Given a space $\mathscr{S}$, real numbers α, β, feasibility predicate feas, and an associated set of explanation functions **EF**, an *explanation function distribution* is a finitary[3] probability distribution $\mathsf{exfd} : \mathbf{EF} \to [0,1]$ with $\sum_{\mathsf{ex\_fcn} \in \mathbf{EF}} \mathsf{exfd}(\mathsf{ex\_fcn}) = 1$. Let **EFD** be a set of explanation function distributions.

We use $|\mathsf{exfd}|$ to denote the cardinality of the set **EF** associated with exfd.

Example 4.6. Following from Example 4.5, we define the explanation function distribution $\mathsf{exfd}_{burglar}$ that assigns a uniform probability to explanation functions in the set $\mathsf{ex\_fcn}_1, \mathsf{ex\_fcn}_2$ (*i.e.*, $\mathsf{exfd}_{burglar}(\mathsf{ex\_fcn}_1) = 0.5$).

We now define an *expected reward* that takes mixed strategies specified by explanation function distributions into account to compute an expected value for the reward function to return.

Definition 4.10 (Expected Reward). Given a reward function **rf**, and explanation function distributions $\mathsf{exfd}_{adv}, \mathsf{exfd}_{ag}$ for the adversary and agent respectively, the *expected reward* is a function $\mathsf{EXR}^{\mathbf{rf}} : \mathbf{EFD} \times \mathbf{EFD} \to [0,1]$. For some explanation function distributions $\mathsf{exfd}_{adv}, \mathsf{exfd}_{ag}$, we define $\mathsf{EXR}^{\mathbf{rf}}(\mathsf{exfd}_{adv}, \mathsf{exfd}_{ag})$ as follows:

$$\sum_{\mathsf{ex\_fcn}_{adv} \in \mathbf{EF}_{adv}} \left(\mathsf{exfd}_{adv}(\mathsf{ex\_fcn}_{adv}) \cdot \sum_{\mathsf{ex\_fcn}_{ag} \in \mathbf{EF}_{ag}} \mathsf{exfd}_{ag}(\mathsf{ex\_fcn}_{ag}) \cdot \mathbf{rf}(\mathsf{ex\_fcn}_{adv}, \mathsf{ex\_fcn}_{ag}) \right)$$

This definition can be explained as follows. We consider each possible explanation function $\mathsf{ex\_fcn}_{ad}$ that the adversary might use. For each possible explanation function $\mathsf{ex\_fcn}_{ag}$ used by the agent, we find the *expected reward* to the agent, which

---

[3] That is, exfd assigns non-zero probabilities to only finitely many explanation functions.

is the probability of the agent using explanation function $\text{exfd}_{ag}(\text{ex\_fcn}_{ag})$ times the reward to the agent if he uses ex_fcn_{ag} and the adversary uses ex_fcn_{adv}—this is the product $\text{exfd}_{ag}(\text{ex\_fcn}_{ag}) \cdot \mathbf{rf}(\text{ex\_fcn}_{adv}, \text{ex\_fcn}_{ag})$ in the above formula. However, this product must be multiplied by the probability that the adversary uses explanation function ex_fcn_{adv}, yielding $\text{exfd}_{adv} \times \text{exfd}_{ag}(\text{ex\_fcn}_{ag}) \cdot \mathbf{rf}(\text{ex\_fcn}_{adv}, \text{ex\_fcn}_{ag})$. This expression is then summed up over all possible explanation functions ex_fcn_{adv} that the adversary might use to give the final expected reward (for the agent).

In this chapter, we will generally not deal with expected reward directly. Rather, we handle two special cases—expected adversarial detriment and expected agent benefit—in which the adversary's and agent's strategies are *not* mixed respectively. We explore these two special cases in the next two sections.

4.3 Selecting a Strategy for the Adversary

In this section, we consider the problem where the adversary has already decided what the set $\mathcal{O}$ of observations should be (*e.g.*, in the case of the insurgents, this would correspond to the insurgents having selected the targets of their terror attacks, while in the case of the burglar, this corresponds to the set of targets the burglar plans to break into), and now he wants to choose the best strategy $\mathscr{A}$ to carry out his nefarious deeds. Of course, the strategy $\mathscr{A}$ needs to be an explanation for $\mathcal{O}$ with respect to a given feasibility predicate. We assume the adversary has a probabilistic model of the agent's behavior (an explanation function distribution) and that he wants to eventually find an explanation (*e.g.*, the set of locations for his weapons caches). Even though he can use "expected reward" to measure how close the agent will be to the adversary's explanation, only the agent's strategy is mixed because the adversary must physically select his strategy once and for all (*e.g.*, the burglar must decide where to live/work, while the terrorist must decide where to place his weapons caches). In other words, the adversary's strategy is not mixed—it is concrete. In order to account for this, we introduce a special case of expected reward called the *expected adversarial detriment* (of a given strategy $\mathscr{A}$ that he chooses).

Definition 4.11 (Expected Adversarial Detriment). Given any reward function $\mathbf{rf}$ and explanation function distribution exfd, the *expected adversarial detriment* is the function $\text{EXD}^{\mathbf{rf}} : \mathbf{EFD} \times 2^{\mathscr{S}} \to [0,1]$ defined as follows:

$$\text{EXD}^{\mathbf{rf}}(\text{exfd}, \mathscr{A}) = \sum_{\text{ex\_fcn} \in \mathbf{EF}} \mathbf{rf}(\mathscr{A}, \text{ex\_fcn}(\mathcal{O}, k)) \cdot \text{exfd}(\text{ex\_fcn})$$

Intuitively, the expected adversarial detriment is the expected number of partner locations the agent may uncover *according to the explanation function distribution* exfd that the adversary uses to model the agent. To compute this, we consider each explanation function ex_fcn and identify the adversary's reward—which we call "detriment", since it is a measure the adversary wishes to minimize. The product of the two gives the expected detriment if in fact the explanation function distribution is correct and the sum of these products, one for each possible explanation function ex_fcn, gives the total expected detriment.

We illustrate the expected adversary detriment via the following example.

Example 4.7. Following from the previous examples, suppose the burglar is planning to have three safe locations (*e.g.*, his house, his office, and his significant other's house). Suppose, from prior experience of the police (or by appropriate scouting), he expects that police detectives will look for his safe houses using $exfd_{burglar}$ (see Example 4.6). One suggestion the burglar may consider is to choose safe houses at locations p_{41}, p_{52} (see Figure 4.1). Note that this explanation is optimal with respect to cardinality. With $dist = 100$ meters, he wishes to compute $EXD^{crf}(exfd_{burglar}, \{p_{41}, p_{52}\})$. We first need to find the reward associated with each explanation function (see Example 4.5):

$$crf^{dist}(\{p_{41}, p_{52}\}, ex\_fcn_1(\mathscr{O}, 3)) = 1$$
$$crf^{dist}(\{p_{41}, p_{52}\}, ex\_fcn_2(\mathscr{O}, 3)) = 0.5$$

Thus, $EXD^{crf}(exfd_{burglar}, \{p_{41}, p_{52}\}) = 0.5 \cdot 1 + 0.5 \cdot 0.5 = 0.75$. Hence, this is probably not the best location for the burglar to position his safe houses with respect to **crf** and **exfd**, as the expected adversarial detriment associated with this set of locations is large.

The expected adversarial detriment is a quantity that the adversary would seek to minimize. This is now defined as an *optimal adversarial strategy* below.

Definition 4.12 (Optimal Adversarial Strategy). Given a set of observations $\mathscr{O}$, natural number k, reward function **rf**, and explanation function distribution **exfd**, an **optimal adversarial strategy** is a k-sized explanation $\mathscr{A}$ for $\mathscr{O}$ such that $EXD^{rf}(exfd, \mathscr{A})$ is minimized.

4.3.1 The Complexity of Finding an Optimal Adversarial Strategy

In this section, we formally define the optimal adversarial strategy (OAS) problem and study its complexity.

OAS Problem
INPUT: Space $\mathscr{S}$, feasibility predicate feas, real numbers α, β, set of observations $\mathscr{O}$, natural number k, reward function **rf**, and explanation function distribution **exfd**.
OUTPUT: The optimal adversarial strategy $\mathscr{A}$.

The result below demonstrates that the known NP-hard problem *Geometric Covering by Discs* [2] is polynomially reducible to OAS. This establishes NP-hardness.

Theorem 4.1. *OAS is NP-hard.*

Proof. CONSTRUCTION: Given an input $\langle P, b, K \rangle$ of GCD, we create an instance of **OAS** in PTIME as follows:

- Set $\mathscr{S}$ to be a grid large enough that all points in P are also points in $\mathscr{S}$.
- feas$(p) =$ TRUE if and only if $p \in P$
- $\alpha = 0$, $\beta = b$, $\mathscr{O} = P$, $k = |P|$
- Let $\mathbf{rf}(\mathscr{E}_1, \mathscr{E}_2) = 1$ if $\mathscr{E}_1 \subseteq \mathscr{E}_2$, and $\frac{|\mathscr{E}_1|}{|\mathscr{S}|}$ otherwise.
  This satisfies reward Axiom 1 as $\mathscr{E}_1 \subseteq \mathscr{S}$, Axiom 2 by definition, and the satisfaction of Axiom 3, along with monotonicity (with respect to the second argument) can easily be shown by the fact that explanations that are not supersets of $\mathscr{E}_1$ (called $\mathscr{E}_2, \mathscr{E}_3$) satisfy $\mathbf{rf}(\mathscr{E}_1, \mathscr{E}_2) = \mathbf{rf}(\mathscr{E}_1, \mathscr{E}_3)$.
- Let ex_fcn(O, num) that returns set O when $num = |O|$ and is otherwise undefined. Let exfd$(\text{ex\_fcn}) = 1$ and 0 otherwise.

CLAIM 1: If $\mathscr{A}$ as returned by **OAS** has a cardinality of $\leq K$, then the answer to GCD is "yes".
Suppose, by way of contradiction, that $card(\mathscr{A}) \leq K$ and GCD answers "no." This is an obvious contradiction as $\mathscr{A}$ is a subset of P (by how feasibility was defined) where all elements of P are within a radius of b and $\mathscr{A}$ also meets the cardinality requirement of GCD.

CLAIM 2: If the answer to GCD is "yes" then $\mathscr{A}$ as returned by **OAS** has a cardinality of less than or equal to K.
Suppose, by way of contradiction, GCD returns "yes" but $\mathscr{A}$ returned by **OAS** has a cardinality greater than K. By the result of GCD, there exists a set P' of cardinality K such that each point in P (hence $\mathscr{O}$) is of a distance $\leq \beta$ from a point in P'. This, along with the definition of feasibility, makes P' a valid K-explanation for $\mathscr{O}$. We note that ex_fcn$(P, |P|) = P$ and that exfd assigns this reward function a probability of one. Hence, the expected adversarial detriment for any explanation $\mathscr{A}'$ is $\mathbf{rf}(\mathscr{A}', P)$. As P' is an explanation of cardinality less than $\mathscr{A}$, it follows that $\mathbf{rf}(P', P) < \mathbf{rf}(\mathscr{A}, P)$, which is a contradiction.

The proof of the above theorem yields two insights, stated below as a corollary.

Corollary 4.1. *1. OAS is NP-hard even if the reward function is monotonic (or antimonotonic).*
*2. OAS is NP-hard even if the cardinality of **EF** is 1.*

Thus, we cannot simply pick an "optimal" function from **EF**. To show an upper bound, we define OAS-DEC to be the decision problem associated with OAS. If the reward function is computable in polynomial time, then the following result says that OAS-DEC is in the complexity class NP.

OAS-DEC
INPUT: Space $\mathscr{S}$, feasibility predicate feas, real numbers α, β, set of observations $\mathscr{O}$, natural number k, reward function $\mathbf{rf}$, explanation function distribution exfd, and number $R \in [0, 1]$.

OUTPUT: "Yes" if there exists an adversarial strategy $\mathscr{A}$ such that $\mathsf{EXD^{rf}}(\mathsf{exfd}, \mathscr{A}) \leq R$, and "no" otherwise.

Theorem 4.2. *If the reward function is computable in PTIME, then OAS-DEC is NP-complete.*

Proof. NP-hardness follows from Theorem 4.1. To show NP-completeness, a witness simply consists of $\mathscr{A}$. We note that, as the reward function is computable in PTIME, finding the expected adversarial detriment for $\mathscr{A}$ and comparing it to R can also be accomplished in PTIME.

Suppose we have an NP oracle that can return an optimal adversarial strategy and suppose this NP oracle returns $\mathscr{A}$. Quite obviously, this is the **best response** of the adversary to the mixed strategy of the agent. Now, how does the agent respond to such a strategy? If we were to assume that such a solution were unique, then the agent would simply have to find a strategy $\mathscr{B}$ such that $\mathbf{rf}(\mathscr{A}, \mathscr{B})$ is maximized. This is a special case of the problem we discuss in Section 4.4. However, this is not necessarily the case. A natural way to address this problem is to create a uniform probability distribution over all optimal adversarial strategies and optimize the expected reward—again a special case of what we will discuss later in Section 4.4. However, obtaining the set of explanations is not an easy task. Even if we had an easy way to exactly compute an optimal adversarial strategy, finding *all* such strategies is an even more challenging problem. In fact, it is at least as hard as the counting version of GCD—which we already have shown to be #P-hard and difficult to approximate (see Chapter 2). The following theorem shows that finding the set of all optimal strategies (for the adversary) that have an expected adversarial detriment below a certain threshold is #P-hard.

Theorem 4.3. *Finding the set of all adversarial optimal strategies that provide a "yes" answer to OAS-DEC is #P-hard.*

Proof. Let us assume that we know one optimal adversarial strategy and can compute the expected adversarial detriment from such a set. Let us call this value D. Given an instance of GCD, we can create an instance of **OAS-DEC** as in Theorem 4.1, where we set $R = D$. Suppose we have an algorithm that produces all adversarial strategies. If we iterate through all strategies in this set, and count all strategies with a cardinality $\leq K$ (the K from the instance of GCD), we have counted all solutions to GCD—thereby solving the counting version of GCD, a #P-hard problem that is difficult to approximate by Lemma 2.1.

This theorem says that it is infeasible for the adversary to find all strategies that are optimal for him.

4.3.2 Pre-Processing and Naive Approach

In this section, we present several algorithms to solve OAS. We first present a simple routine for pre-processing followed by a naive enumeration-based algorithm.

We use Δ to denote the maximum number of partners per observation and f to denote the maximum number of observations supported by a single partner. In general, Δ is bounded by $\pi(\beta^2 - \alpha^2)$, but may be lower depending on the feasible points in $\mathscr{S}$. Likewise, f is bounded by $\min(|\mathcal{O}|, \Delta)$ but may be much smaller depending on the sparseness of the observations.

Pre-Processing Procedure. Given a space $\mathscr{S}$, a feasibility predicate feas, real numbers $\alpha, \beta \in [0,1]$, and a set $\mathcal{O}$ of observations, we create two lists (similar to a standard inverted index) as follows.

- **Matrix M.** M is an array of size $\mathscr{S}$. For each point $p \in \mathscr{S}$, $M[p]$ is a *list of pointers* to observations. $M[p]$ contains pointers to each observation o such that feas(p) is true and such that $d(o,p) \in [\alpha, \beta]$.
- **List L.** List L contains a pointer to position $M[p]$ in the array M if and only if there exists an observation $o \in \mathcal{O}$ such that feas(p) is true and such that $d(o,p) \in [\alpha, \beta]$.

Thus, $M[p]$ points to all observations that are both feasible and which are within the appropriate lower and upper bounds $[\alpha, \beta]$ in distance from point p. In the case of the insurgent cache detection problem, $M[p]$ specifies the set of attacks that the insurgent terror group wants to carry that could be carried out if the insurgent group had a weapons cache at location p. In contrast, list L points to all locations that can be used to carry out at least one of the attacks that the insurgent group wants to carry out. What is important to note is that if a point p is *not* in L, then point p is not a location where the adversary might want to put his weapons cache.

It is easy to see that we can compute M and L in $O(|\mathcal{O}| \cdot \Delta)$ time. The example below shows how M, L apply to our running burglary example.

Example 4.8. Consider our running example concerning the burglaries and the burglar's dwellings that started with Example 4.1. The set L consists of $\{p_1, \ldots, p_{67}\}$. The matrix M returns lists of observations that can be associated with each point. For example, $M(p_{40}) = \{o_3, o_4, o_5\}$ and $M(p_{46}) = \{o_1, o_2\}$.

Naive Approach. After pre-processing, a straightforward exact solution to OAS would be to enumerate all subsets of L that have a cardinality less than or equal to k. Let us call this set L^*. Furthermore, suppose we eliminate all elements of L^* that are not valid explanations. Finally, for each element of L^*, suppose we compute the expected adversarial detriment and return the element of L^* for which this value is the least. Clearly, this approach is impractical as the cardinality of L^* can be very large. Furthermore, this approach does not take advantage of the specific reward functions. We now present mixed integer programs (MIPs) to compute the minimal expected adversary detriment when the associated reward function is either **wrf** or **frf**. We first write these mixed integer programs—later, we develop methods to reduce the complexity of solving these MILPs.

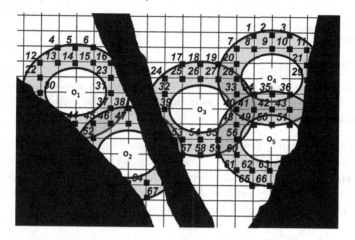

Fig. 4.3 Set L of all possible partners for our burglar's dwelling locations.

4.3.3 Mixed Integer Programs for OAS under wrf, crf, frf

We present mixed integer linear programs (MILPs) to solve OAS exactly for some specific reward functions. First, we present a mixed integer linear program for the reward function **wrf**. Later, in Section 4.3.4, we show how to improve efficiency—while maintaining optimality—by reducing the number of variables in the MILP. Note that these constraints can also be used for **crf** as **wrf** generalizes **crf**. We also define a MILP for the **frf** reward function.

While these mixed integer programs may appear nonlinear, Proposition 4.9 gives a simple transformation to standard linear form. For readability, we define the MILPs before discussing this transformation.

We start by associating an integer-valued variable X_i with each point $p_i \in L$. Intuitively, X_i is an (unknown) variable which has the value 0 if the adversary chooses not to locate a partner location there, and 1 otherwise.

Definition 4.13 (wrf MILP). Given real number $dist > 0$ and weight function W, associate a constant w_i with the weight $W(p_i)$ of each point $p_i \in L$. Next, for each $p_i \in L$ and $\mathsf{ex\_fcn}_j \in \mathbf{EF}$, let constant $c_{i,j} = 1$ if and only if $\exists p' \in \mathsf{ex\_fcn}(\mathcal{O}, k)$ such that $d(p', p_i) \leq dist$ and 0 otherwise. Finally, associate an integer-valued variable X_i with each $p_i \in L$.
Minimize:

$$\sum_{\mathsf{ex\_fcn}_j \in \mathbf{EF}} \left(\mathsf{exfd}(\mathsf{ex\_fcn}_j) \cdot \sum_{p_i \in L} \left(X_i \cdot \frac{w_i \cdot c_{i,j}}{\sum_{p_i \in L} w_i \cdot X_i} \right) \right)$$

subject to:

1. $X_i \in \{0, 1\}$

2. Constraint $\sum_{p_i \in L} X_i \leq k$
3. For each $o_j \in \mathcal{O}$, add constraint
$$\sum_{p_i \in L_{d(o_j, p_i) \in [\alpha, \beta]}} X_i \geq 1$$

We note that the above definition does not define purely linear constraints. Fortunately, these constraints can be easily linearized using the well-known Charnes-Cooper transformation [3].

Example 4.9. Continuing from Examples 4.7 (page 109) and 4.8, suppose the burglar wishes to produce an adversarial strategy $\mathscr{A}$ using **wrf**. Consider the case where we use **crf**, $k \leq 3$, and *dist* = 100 meters as before (see Example 4.7). Clearly, there are 67 variables in these constraints, as this is the cardinality of set L (as per Example 4.8). The constants $c_{i,1}$ are 1 for elements in the set:

$$\{p_{35}, p_{40}, p_{41}, p_{42}, p_{43}, p_{44}, p_{45}, p_{46}, p_{49}, p_{49}, p_{50}, p_{52}, p_{56}\}$$

(and 0 for all others). The constants $c_{i,2}$ are 1 for elements in the set

$$\{p_{33}, p_{37}, p_{40}, p_{41}, p_{45}, p_{46}, p_{47}, p_{48}\}$$

(and 0 for all others).

As in the case of the weighted reward function above, we can create a MILP for the falloff reward function **frf** as follows, where X_i has a meaning identical to that in the preceding case.

Definition 4.14 (frf MILP). For each $p_i \in L$ and ex_fcn$_j \in$ **EF**, let constant $c_{i,j} = \min_{p' \in \text{ex\_fcn}(\mathcal{O}, k)}(d(p_i, p')^2)$. Associate an integer-valued variable X_i with each $p_i \in L$.
Minimize:

$$\sum_{\text{ex\_fcn}_j \in \text{EF}} \left(\text{exfd}(\text{ex\_fcn}_j) \cdot \sum_{p_i \in L} \left(X_i \cdot \frac{1}{c_{i,j} + \sum_{p_i \in L} X_i} \right) \right)$$

subject to:

1. $X_i \in \{0, 1\}$
2. Constraint $\sum_{p_i \in L} X_i \leq k$
3. For each $o_j \in \mathcal{O}$, add constraint
$$\sum_{p_i \in L_{d(o_j, p_i) \in [\alpha, \beta]}} X_i \geq 1$$

The following theorem tells us that solving the above MILPs correctly yields a solution for the OAS problem under both **wrf** or **frf**.

Proposition 4.6. *Suppose $\mathscr{S}$ is a space, $\mathcal{O}$ is an observation set, real numbers $0 \leq \alpha < \beta \leq 1$, and suppose the **wrf** and **frf** MILPs are defined as above.*

1. *Suppose $\mathscr{A} = \{p_1, \ldots, p_n\}$ is a solution to OAS with* **wrf** *(resp.* **frf***). Consider the assignment that assigns 1 to each $X_1, \ldots, X_n$ corresponding to the p_i's and 0 otherwise. This assignment is an optimal solution to the MILP.*
2. *Given the solution to the constraints, if for every $X_i = 1$, we add point p_i to set $\mathscr{A}$, then $\mathscr{A}$ is a solution to OAS with* **wrf** *(resp.* **frf***).*

Proof. PART 1: Suppose, by way of contradiction, that there is a set of variables $X'_1, \ldots, X'_m$ that is a solution to the constraints such that the value of the objective function is less than if variables $X_1, \ldots, X_n$ were used. Then, there are points $p'_1, \ldots, p'_m$ in set L that correspond with the X_i's such that they cover all observations and that the expected adversarial detriment is minimized. Clearly, this is a contradiction.

PART 2: Suppose, by way of contradiction, that there is a set of points $\mathscr{A}'$ such that the expected adversarial detriment is less than $\mathscr{A}$. Clearly, $\mathscr{A}$ is a valid explanation that minimizes the expected adversarial detriment by the definition of the constraints—hence a contradiction.

The result below states that setting up either set of constraints can be performed in polynomial time, where computing the $c_{i,j}$ constants is the dominant operation.

Proposition 4.7. *Setting up the* **wrf/frf** *constraints can be accomplished in $O(|\mathbf{EF}| \cdot k \cdot |\mathscr{O}| \cdot \Delta)$ time (provided the weight function W can be computed in constant time).*

Proof. First, we must run POSS-PART, which requires $O(|\mathscr{O}| \cdot \Delta)$ operations. This results in a list of size $O(|\mathscr{O}| \cdot \Delta)$. For each explanation function, ex_fcn, we must compare every element in L with each element of ex_fcn($\mathscr{O}$), which would require $O(k \cdot |\mathscr{O}| \cdot \Delta)$ time. As there are $|\mathbf{EF}|$ explanation functions, the statement follows.

The number of variables for either set of constraints is related to the size of L, which depends on the number of observations, spacing of $\mathscr{S}$, and α, β.

Proposition 4.8. *The* **wrf/frf** *constraints have $O(|\mathscr{O}| \cdot \Delta)$ variables and $1 + |\mathscr{O}|$ constraints.*

Proof. As list L is of size $O(|\mathscr{O}| \cdot \Delta)$, and there is one variable for every element of L, there are $O(|\mathscr{O}| \cdot \Delta)$ variables. As there is a constraint for each observation, plus a constraint to ensure the cardinality requirement (k) is met, there are $1 + |\mathscr{O}|$ constraints.

The MILPs for **wrf** and **frf** appear nonlinear as the objective function is fractional. However, as the denominator is non-zero and strictly positive, the Charnes-Cooper transformation [3] allows us to quickly (in the order of number of constraints multiplied by the number of variables) transform the constraints into a purely integer-linear form. Many linear and integer-linear program solvers include this transformation in their implementation and hence, this transformation is very standard.

Proposition 4.9. *The* **wrf/frf** *constraints can be transformed into a purely linear-integer form in $O(|\mathscr{O}|^2 \cdot \Delta)$ time.*

Proof. Obviously, in both sets of constraints, the denominator of the objective function is strictly positive and non-zero. Hence, we can directly apply the Charnes-Cooper transformation [3] to obtain a purely integer-linear form. This transformation requires $O(number\ of\ variables \times number\ of\ constraints)$. Hence, the $O(|\mathcal{O}|^2 \cdot \Delta)$ time complexity of the operation follows immediately from Proposition 4.8.

We note that a linear relaxation of any of the above three constraints can yield a lower bound on the objective function in $O(|L|^{3.5})$ time.

Proposition 4.10. *Given the constraints of Definition 4.13 or Definition 4.14, if we consider the linear program formed by setting all X_i variables to be in $[0,1]$, then the value returned by the objective function will be a lower bound on the value returned by the objective function for the mixed integer-linear constraints, and this value can be obtained in $O(|\mathcal{O}|^{3.5} \cdot \Delta^{3.5})$ time.*

Proof. CLAIM 1: The linear relaxation of Definition 4.13 or Definition 4.14 provides a lower bound on the objective function value for the full integer-linear constraints. As an optimal value returned by the integer-linear constraints would also be a solution, optimal with respect to minimality, for the linear relaxation, the statement follows.
CLAIM 2: The lower bound can be obtained in $O(|L|^{3.5})$ time.
As there is a variable for each element of L, the size of L is $O(|\mathcal{O}| \cdot \Delta)$, and the claim follows immediately from the result of [4].

Likewise, if we solve the mixed integer linear program with a reduced number of variables, we are guaranteed that the solution will cause the objective function to be an upper bound for the original set of constraints.

Proposition 4.11. *Consider the MILPs in Definition 4.13 and Definition 4.14. Suppose $L' \subset L$ and every variable X_i associated with some $p_i \in L'$ is set to 0. The resulting solution is an upper bound on the objective function for the constraints solved on the full set of variables.*

Proof. Suppose, by way of contradiction, that the solution for the objective function on the reduced MILP would be less than the actual MILP. Let $X_1, \ldots, X_n$ be the variables set to 1 for the reduced MILP in this scenario. We note, that setting the same variables to the full MILP would also be a solution, and could not possibly be less than a minimal solution. This is a contradiction.

4.3.4 Correctly Reducing the Number of Variables for crf

As the complexity of solving MILPs is closely related to the number of variables in the MILP, the goal of this section is to reduce the number of variables in the MILP associated above with the **crf** reward function. In this section, *we show that if we can find a certain type of explanation called a δ-core optimal explanation, then we can* *"build up"* an optimal adversarial strategy in polynomial time.[4] It also turns out that finding these special explanations can be accomplished using a MILP which will often have significantly less variables than the MILP's of the last section. First, we consider the **wrf** constraints applied to **crf** which is a special case of **wrf**. The objective function for this case is:

$$\sum_{\text{ex\_fcn}_j \in \mathbf{EF}} \left(\text{exfd}(\text{ex\_fcn}_j) \cdot \sum_{p_i \in L} \left(X_i \cdot \frac{c_{i,j}}{\sum_{p_i \in L} X_i} \right) \right)$$

where for each $p_i \in L$ and $\text{ex\_fcn}_j \in \mathbf{EF}$, $c_{i,j} = 1$ if and only if $\exists p' \in \text{ex\_fcn}_j(\mathscr{O}, k)$ such that $d(p', p_i) \leq dist$ and 0 otherwise. If we rearrange the objective function, we see that with each X_i variable associated with point $p_i \in L$, there is an associated constant $const_i$:

$$const_i = \sum_{\text{ex\_fcn}_j \in \mathbf{EF}} \text{exfd}(\text{ex\_fcn}_j) \cdot c_{i,j}.$$

This lets us rewrite the objective function as:

$$\frac{\sum_{p_i \in L} X_i \cdot const_i}{\sum_{p_i \in L} X_i}.$$

Example 4.10. Continuing from Example 4.9, $const_i = 0.5$ for the following elements: $\{p_{33}, p_{35}, p_{37}, p_{42}, p_{43}, p_{44}, p_{47}, p_{49}, p_{50}, p_{52}, p_{56}\}$; $const_i = 1$ for these elements: $\{p_{40}, p_{41}, p_{45}, p_{46}, p_{48}\}$, and 0 for all others.

4.3.4.1 Relationship with Covering Problems

In many covering problems where we wish to find a cover of minimal cardinality, we could reduce the number of variables in the integer program by considering equivalent covers as duplicate variables. However, for OAS, this technique can not be easily applied. The reason for this is because an optimal adversarial explanation is not necessarily irredundant (see Definition 2.7, page 24). Consider the following. Suppose we wish to find an optimal adversarial strategy of size k. Let P be an irredundant cover of size $k - 1$. Suppose there is some element $p' \in P$ that covers only one observation o'. Hence, there is no $p \in P - \{p'\}$ that covers o' by the definition of an irredundant cover. Suppose there is also some $p'' \notin P$ that also covers o'. Now, let $m = \sum_{p_i \in P - \{p'\}} const_i$. In our construction of an example solution to OAS that is not irredundant, we let $const'$ be the value associated with both p' and p''. Consider the scenario where $const' < \frac{m}{k-2}$. Suppose by way of contradiction that the optimal irredundant cover is also the optimal adversarial strategy. Then, by the definition of an optimal adversarial strategy we know that the set P is more optimal than $P \cup \{p''\}$. This would mean that $\frac{m+const'}{k-1} < \frac{m+2 \cdot const'}{k}$. This leads us to infer that

---

[4] Thus, this describes a class of OAS problems that can be solved exactly in polynomial time.

$m < const' \cdot (k-2)$, which clearly contradicts $const' < \frac{m}{k-2}$. It is clear that a solution to OAS need not be irredundant.

However, we do leverage the idea of an irredundant cover in a different exact approach in this section which may provide a speedup over the exact algorithms of the previous section. The main intuition is that each OAS solution contains an irredundant cover, and if we find such a cover, we can build an optimal adversarial strategy in polynomial time. First, we define a *core* explanation. Before doing so, we recall that L is the set of all points in the space $\mathscr{S}$ that are feasible and that explain at least one observation (*i.e.*, is within the $[\alpha, \beta]$ distance bounds from at least one explanation).

Definition 4.15 (Core Explanation). Given an observation set $\mathscr{O}$ and set L of possible partners, an explanation $\mathscr{E}_{core}$ is a **core explanation** if and only if for any $p_i \in \mathscr{E}_{core}$, there does not exist $p_j \in L$ such that:

1. $\forall o \in \mathscr{O}$ if o, p_i are partners, then o, p_j are also partners.
2. $const_j < const_i$

We now show that any optimal adversarial strategy contains a subset that is a core explanation.

Theorem 4.4. *If $\mathscr{A}$ is an optimal adversarial strategy, there exists a core explanation $\mathscr{E}_{core} \subseteq \mathscr{A}$.*

Proof. CLAIM 1: For any explanation $\mathscr{E}$, there is an explanation $\mathscr{E}' \subseteq \mathscr{E}$ such that there are no two elements $p, p' \in \mathscr{E}'$ such that $\forall o \in \mathscr{O}$ such that o, p are partners, then o, p' are also partners.
Consider $\mathscr{E}$. If it does not already have the quality of Claim 1, then by simple induction, we can remove elements until the resulting set does.
CLAIM 2: If $\mathscr{A}$ is an optimal adversarial strategy, there is a no $p_j \in L - \mathscr{A}$ such that there exists $p_i \in \mathscr{A}$ where $const_j < const_i$ and $\forall o \in \mathscr{O}$ such that o, p_i are partners, then o, p_j are also partners.
Suppose, by way of contradiction, there is a $p_j \in L - \mathscr{A}$ such that there exists $p_i \in \mathscr{A}$ where $const_j < const_i$ and $\forall o \in \mathscr{O}$ such that o, p_i are partners, then o, p_j are also partners. Consider the set $(\mathscr{A} - \{p_i\} \cup \{p_j\})$. This set is still an explanation and $\mathsf{EXD^{rf}}(\mathsf{exfd}, (\mathscr{A} - \{p_i\} \cup p_j) < \mathsf{EXD^{rf}}(\mathsf{exfd}, \mathscr{A})$—which would be a contradiction as $\mathscr{A}$ is an optimal adversarial strategy.
CLAIM 3: There is an explanation $\mathscr{E} \subseteq \mathscr{A}$ such that condition 1 of Definition 4.15 holds.
Consider the set $\mathscr{E} = \{p_i \in \mathscr{A} \mid \not\exists p_j \in \mathscr{A} \ s.t. \ (const_j < const_i) \wedge$
$(\forall o \in \mathscr{O} \ s.t. \ o, p_i \ are \ partners, \ then \ o, p_j \ are \ also \ partners)\}$. By Claim 1, this set is contained in an OAS. Note that any observation covered by a point in $\mathscr{A} - \mathscr{E}$ is covered by a point in $\mathscr{E}$, so $\mathscr{E}$ is an explanation. Further, by the definition of $\mathscr{E}$ and Claim 2, this set meets condition 1 of Definition 4.15.
CLAIM 4: Set $\mathscr{E}$ from Claim 3 is a core explanation.
By Claim 3, $\mathscr{E}$ is a valid explanation and meets condition 1 of Definition 4.15.

Example 4.11. Continuing from Example 4.10, consider the set $\mathscr{A} = \{p_{34}, p_{38}, p_{57}\}$ (which would correspond to drug lab locations as planned by the cartel). Later, we show that this is an optimal adversarial strategy (the expected adversarial detriment associated with $\mathscr{A}$ is 0). Consider the subset p_{34}, p_{38}. As p_{34} explains observations o_3, o_4, o_5 and p_{38} explains observations o_1, o_2, this set is also an explanation. Obviously, it is of minimal cardinality. Hence, the set $\{p_{34}, p_{38}\}$ is a **core explanation** of $\mathscr{A}$.

Suppose we have an oracle that, for a given k, $\mathscr{O}$, and exfd returns a core explanation $\mathscr{E}_{core}$ that is guaranteed to be a subset of the optimal adversarial strategy associated with k, $\mathscr{O}$, and exfd. The following theorem says we can find the optimal adversarial strategy in polynomial time. The key intuition is that we need not concern ourselves with covering the observations as $\mathscr{E}_{core}$ is an explanation. The algorithm BUILD-STRAT follows from this theorem.

Theorem 4.5. *If there is an oracle that for any given k, $\mathscr{O}$, and exfd returns a core explanation $\mathscr{E}_{core}$ that is guaranteed to be a subset of the optimal adversarial strategy associated with k, $\mathscr{O}$, and exfd, then we can find an optimal adversarial strategy in $O(\Delta \cdot |\mathscr{O}| \cdot \log(\Delta \cdot |\mathscr{O}|) + (k - |\mathscr{E}_{core}|)^2)$ time.*

Proof. CLAIM 1: For explanation $\mathscr{E}$ and point $p_i \in L - \mathscr{E}$, $\text{EXD}^{\text{rf}}(\text{exfd}, \mathscr{E}) > \text{EXD}^{\text{rf}}(\text{exfd}, \mathscr{E} \cup \{p_i\})$ if and only if $const_i < \text{EXD}^{\text{rf}}(\text{exfd}, \mathscr{E})$.
If: Suppose $const_i < \text{EXD}^{\text{rf}}(\text{exfd}, \mathscr{E})$. Let $\text{EXD}^{\text{rf}}(\text{exfd}, \mathscr{E}) = \frac{a}{b}$. Hence, $\text{EXD}^{\text{rf}}(\text{exfd}, \mathscr{E} \cup \{p_i\}) = \frac{a + const_i}{b + 1}$. Suppose, by way of contradiction, $\text{EXD}^{\text{rf}}(\text{exfd}, \mathscr{E}) \leq \text{EXD}^{\text{rf}}(\text{exfd}, \mathscr{E} \cup \{p_i\})$. Then, $\frac{a}{b} \leq \frac{a + const_i}{b + 1}$. This give us $a \cdot b + a \leq a \cdot b + const_i \cdot b$, which give us $\text{EXD}^{\text{rf}}(\text{exfd}, \mathscr{E}) \leq const_i$—a contradiction.
Only-if: Suppose $\text{EXD}^{\text{rf}}(\text{exfd}, \mathscr{E}) > \text{EXD}^{\text{rf}}(\text{exfd}, \mathscr{E} \cup \{p_i\})$. Let $\text{EXD}^{\text{rf}}(\text{exfd}, \mathscr{E}) = \frac{a}{b}$. Hence, $\frac{a}{b} > \frac{a + const_i}{b + 1}$, which proves the claim.
CLAIM 2: For explanation $\mathscr{E}$ and points $p_i, p_j \in L - \mathscr{E}$ if $const_i < const_j$, then $\text{EXD}^{\text{rf}}(\text{exfd}, \mathscr{E} \cup \{p_i\}) > \text{EXD}^{\text{rf}}(\text{exfd}, \mathscr{E} \cup \{p_j\})$.
Straightforward algebra similar to Claim 1.
CLAIM 3: Algorithm BUILD-STRAT returns an optimal adversarial strategy.
We know that $\mathscr{E}_{core}$ must be in the optimal adversarial strategy. Hence, we suppose BWOC that for the remaining elements there is a better set of elements—cardinality between 0 and $k - |\mathscr{E}_{core}|$ such that the expected adversarial detriment is lower. However, this contradicts Claims 1–2.
CLAIM 4: Algorithm BUILD-STRAT runs in time $O(\Delta \cdot |\mathscr{O}| \cdot \log(\Delta \cdot |\mathscr{O}|) + (k - |\mathscr{E}_{core}|)^2)$.
Sorting the set $L - \mathscr{E}_{core}$ can be accomplished in $O(\Delta \cdot |\mathscr{O}| \cdot \log(\Delta \cdot |\mathscr{O}|))$ time. The remainder can be accomplished in $O((k - |\mathscr{E}_{core}|)^2)$ time.

We now introduce the notion of δ-core optimal. Intuitively, this is a core explanation of cardinality exactly δ that is optimal w.r.t. expected adversarial detriment compared to all other core explanations of that cardinality.

Definition 4.16. Given an integer $\delta > 0$, an explanation distribution function exfd, and a reward function **rf**, a core explanation $\mathscr{E}_{core}$ is δ-**core optimal** if and only if:

Algorithm 11 BUILD-STRAT

INPUT: Partner list L, core explanation $\mathcal{E}_{core}$, natural number k, explanation function distribution exfd

OUTPUT: Optimal adversarial strategy $\mathcal{A}$

1. If $|\mathcal{E}_{core}| = k$, return $\mathcal{E}_{core}$
2. Set $\mathcal{A} = \mathcal{E}_{core}$. Let $k' = |\mathcal{E}_{core}|$
3. Sort the set $L - \mathcal{E}_{core}$ by $const_i$. Let $L' = \{p_1, \ldots, p_{k-k'}\}$ be the $k - k'$ elements of this set with the lowest values for $const_i$, in ascending order
4. For each $p_i \in L'$ let P_i be the set $\{p_1, \ldots, p_i\}$
5. For each P_i let $S_i = \sum_{j \le i} const_j$
6. Let $ans = \min_{p_i \in L'} (\{\frac{k' \cdot \mathsf{EXD^{rf}}(\mathsf{exfd}, \mathcal{E}_{core}) + S_i}{k' + i}\})$
7. Let P_{ans} be the P_i associated with ans
8. If $ans \ge \mathsf{EXD^{rf}}(\mathsf{exfd}, \mathcal{E}_{core})$, return $\mathcal{E}_{core}$, else return $\mathcal{E}_{core} \cup P_{ans}$

- $|\mathcal{E}_{core}| = \delta$
- There does not exist another core explanation $\mathcal{E}'_{core}$ of cardinality exactly δ such that $\mathsf{EXD^{rf}}(\mathsf{exfd}, \mathcal{E}'_{core}) < \mathsf{EXD^{rf}}(\mathsf{exfd}, \mathcal{E}_{core})$

We now define some subsets of the set L that are guaranteed to contain core explanations and δ-core optimal explanations as well. In practice, these sets will be much smaller than L and will be used to create a MILP of reduced size.

Definition 4.17 (Reduced Partner Set). Given observations $\mathcal{O}$ and set of possible partners L, we define the reduced partner set L^{**} as follows:

$$L^{**} = \{p_i \in L \mid \not\exists p_j \in L \; s.t. \; (const_j < const_i) \wedge (\forall o \in \mathcal{O} \; s.t. \; o, p_i \; are \; partners,$$
$$o, p_j \; are \; also \; partners)\}$$

We define L^* as follows:

$$L^* = \{p_i \in L^{**} \mid \not\exists p_j \in L^{**} \; s.t. \; (const_j = const_i) \wedge (\forall o \in \mathcal{O} \; s.t. \; o, p_i \; are \; partners,$$
$$o, p_j \; are \; also \; partners)\}$$

Lemma 4.1. *1. If explanation $\mathcal{E}$ is a core explanation, then $\mathcal{E} \subseteq L^{**}$.*
2. If explanation $\mathcal{E}$ is δ-core optimal, then $\mathcal{E} \subseteq L^{**}$.
3. If for some natural number δ, there exists an explanation of size δ, then there exists a δ-core optimal explanation $\mathcal{E}$ such that $\mathcal{E} \subseteq L^*$.

Proof. Proof of Part 1:
Suppose, BWOC, $\mathcal{E}$ is a core explanation and $\mathcal{E} \not\subseteq L^{**}$. Then, there is some element $p_i \in \mathcal{E} \cap (L - L^{**})$. Moreover, by the definition of a core explanation, there does not exist $p_j \in L$ such that $\forall o \in \mathcal{O}$ such that o, p_i are partners, then o, p_j are also partners and $const_j < const_i$. This would also put the element in L^{**} by the definition of that set—which is a contradiction.

Proof of Part 2:

Suppose, BWOC, there exists explanation $\mathscr{E}$ such that for some δ, $\mathscr{E}$ is δ-core optimal and $\mathscr{E} \not\subseteq L^{**}$. Then, there exists some $p_i \in \mathscr{E} \cap (L - L^{**})$. By the definition of L^{**}, there exists a $p_j \in L^{**}$ such that $const_j < const_i$ and $\forall o \in \mathscr{O}$ s.t. o, p_i are partners, then o, p_j are also partners. Hence, the set $(\mathscr{E} - \{p_i\}) \cup \{p_j\}$ is also an explanation of size δ and has a lower expected detriment. From the definition of δ-core optimal, this is a contradiction.

Proof of Part 3:

Suppose, BWOC, for some δ such that there is an explanation of this size, there does not exist a δ-core optimal explanation $\mathscr{E}$ such that $\mathscr{E} \subseteq L^*$. By the proof of Part 2, we know that a δ-core optimal explanation must be within L^{**}. Further, by the definition of L^*, for any point $p_i \in L^{**} - L^*$, there exists point $p_j \in L^*$ such that $const_j = const_i$ and $\forall o \in \mathscr{O}$ s.t. o, p_i are partners, o, p_j are also partners. Hence, for some δ-core explanation that is not a subset of L^*, any $p_i \in \mathscr{E} \cap (L^{**} - L^*)$ can be replaced by some $p_j \in L^*$, and the resulting set is still an explanation, optimal, and of cardinality δ—a contradiction.

The reduced partner set can be computed in polynomial time. We also note that under the assumption that $|\mathscr{O}| \ll |L|$, which we have found to be true in practice, determining the set L^{**} or L^* can be accomplished faster (in terms of time complexity) than solving even a relaxation of the original MILP.

Proposition 4.12. *Given set L, set L^* and L^{**} can be found in $O(|L|^2 \cdot |\mathscr{O}|^2)$ time.*

Proof. Given sets $L, \mathscr{O}$, set L^{**} can be found with the following steps.

1. For each $p_i \in L$, let $\mathscr{O}_i$ be the subset of $\mathscr{O}$ that can be partnered with p_i
2. For each $p_i \in L$, let $elim_i$ be a boolean variable set to *FALSE*
3. For each $p_i \in L^{**}$, do the following

   a. If not $elim_i$
      i. For each $p_j \in L^{**} - \{p_j\}$, if $\mathscr{O}_j \subseteq \mathscr{O}_i$ and $const_i < const_j$ then set $elim_j = TRUE$

4. Return the set $\{p_i \in L | elim_i = FALSE\}$.

Clearly, the correctness of the above procedure follows directly from the definition of set L^{**}. Further, the complexity of the operation is $O(|L|^2 \cdot |\mathscr{O}|^2)$, as we have two nested loops, each iterating at most $|L|$ times and a comparison where for some p_i, p_j, we examine the elements of $\mathscr{O}_i, \mathscr{O}_j$. To determine the set L^*, we can simply adjust line 3(a)i of the above procedure and change the $<$ to a $\leq$. The correctness again follows from the definition and the time complexity remains the same.

Example 4.12. Let us continue from Example 4.11. Based on pre-processing and the computation of $const_i$, we can easily produce the data of Table 4.1 in polynomial time. Based on this, we obtain a **reduced partner set** $L^* = \{p_{34}, p_{38}, p_{57}\}$.

Next, the following lemma tells us that an OAS must contain a core explanation that is δ-core optimal.

Supported Observations	$const_i = 0$	$const_i = 0.5$	$const_i = 1$
o_1	$p_4 - p_6, p_{12} - p_{16}, p_{22} - p_{23}, p_{30} - p_{31}$	p_{44}	
o_1, o_2	p_{38}	p_{37}, p_{52}	p_{45}, p_{46}
o_2	p_{64}, p_{67}	p_{47}	
o_2, o_3	p_{57}		
o_3	$p_{17} - p_{19}, p_{24} - p_{26}, p_{32}, p_{39}, p_{58} - p_{59}$		
o_3, o_4	$p_{27} - p_{28}$	p_{33}	
o_4	$p_1 - p_3, p_7 - p_{11}, p_{20} - p_{21}, p_{29}, p_{51}$	p_{50}	
o_3, o_4, o_5	$p_{34}, p_{53} - p_{54}$	p_{49}	$p_{40} - p_{41}$
o_5	$p_{36}, p_{60} - p_{66}$	p_{35}	
o_4, o_5		$p_{42} - p_{43}$	
o_3, o_5	p_{55}	p_{56}	p_{48}

Table 4.1 The set L partitioned by $const_i$ and supported observations.

Lemma 4.2. *Given an optimal adversarial strategy $\mathscr{A}$, there exists some $\delta \leq |\mathscr{A}|$ such that there is a δ-core optimal explanation that is a subset of $\mathscr{A}$ (using the **crf** reward function).*

Proof. By Theorem 4.4, $\mathscr{A}$ must contain a core explanation and by Lemma 4.1, any core explanation must be a subset of L^{**}. Therefore, $\mathscr{A} \cap L^{**}$ is a core explanation. Let $B = \mathscr{A} - (\mathscr{A} \cap L^{**})$ and $\delta = |\mathscr{A} \cap L^{**}|$. Suppose $\mathscr{A} \cap L^{**}$ is not δ-core optimal. Then there is some set Q that is a subset of L^{**}, is disjoint from $\mathscr{A} \cap L^{**}$, and is δ-core optimal. Note that $Q \cap B = \emptyset$ as Q must be a subset of L^{**} and B is not. Hence, since it has a lower expected detriment than $\mathscr{A} \cap L^{**}$ and $|Q \cup B| = |\mathscr{A}|$, the set $Q \cup B$ will have a lower expected detriment than $\mathscr{A}$—which is clearly a contradiction as $\mathscr{A}$ is an OAS.

Thus, if we can find the δ-core optimal explanation that is contained in an OAS, we can then find the OAS. If we know δ, such an explanation can be found using a MILP. We now present a set of integer-linear constraints to find a δ-**core optimal** explanation. Of course we can easily adopt the constraints of the previous section, but this would offer us no improvement in performance. We therefore create a MILP that should have a significantly smaller number of variables in most cases.

To create this MILP, we take a given set of possible partners L and calculate the set L^*—the reduced partner set—which often will have a cardinality much smaller than L. Next, we use L^* to form a new set of constraints to find a δ-core optimal explanation. We now present these δ-core constraints. Notice that the cardinality requirement in these constraints is "$=$" and not "$\leq$". This is because Lemma 4.2 ensures a core explanation that is δ-core optimal, meaning that the core explanation must have cardinality exactly δ. This also allows us to eliminate variables from the denominator of the objective function, as the denominator must equal δ as well.

Definition 4.18 (δ-core MILP). Given parameter δ and reduced partner set L^*, we define the δ-core constraints by first associating a variable X_i with each point $p_i \in L^*$, then solving:
Minimize:

$$\frac{1}{\delta} \sum_{p_i \in L^*} X_i \cdot const_i$$

subject to:

1. $X_i \in \{0, 1\}$
2. Constraint $\sum_{p_i \in L} X_i = \delta$
3. For each $o_j \in \mathscr{O}$, add constraint
$$\sum_{p_i \in L^* d(o_j, p_i) \in [\alpha, \beta]} X_i \geq 1$$

Example 4.13. Using set L^* from Example 4.12, we can create δ-core constraints as follows:
Minimize:

$$\frac{1}{\delta} (X_{34} \cdot const_{34} + X_{38} \cdot const_{38} + X_{57} \cdot const_{57})$$

subject to:

1. $X_{34}, X_{38}, X_{57} \in \{0, 1\}$
2. $X_{34} + X_{38} + X_{57} = \delta$
3. $X_{38} \geq 1$ (for observation o_1)
4. $X_{38} + X_{57} \geq 1$ (for observation o_2)
5. $X_{34} + X_{57} \geq 1$ (for observation o_3)
6. $X_{34} \geq 1$ (for observations o_4, o_5)

In the worst case, the set $L^* = L$. Hence, we can assert that:

Proposition 4.13. *The δ-core constraints require $O(\Delta \cdot |\mathscr{O}|)$ variables and $1 + |\mathscr{O}|$ constraints.*

Proof. Mirrors proposition 4.6.

Proposition 4.14. *Given δ-core constraints:*

1. *Given set δ-core optimal explanation $\mathscr{E}_{core} = \{p_1, \ldots, p_n\}$, if variables $X_1, \ldots, X_n$—corresponding with elements in $\mathscr{A}$—are set to 1 and the rest of the variables are set to 0, the objective function of the constraints will be minimized.*
2. *Given the solution to the constraints, if for every $X_i = 1$, we add point p_i to set $\mathscr{E}_{core}$, then $\mathscr{E}_{core}$ is a δ-core optimal solution.*

Proof. From Lemma 4.1, we know that for any δ such that there exists and explanation of that size, there is a δ-core explanation $\mathscr{E}$ that is a subset of L^*. Hence, the rest of the proof mirrors the proof of Proposition 4.6

We now have all the pieces required to leverage core explanations and reduced partner sets to find an optimal adversarial strategy. By Theorem 4.11, we know that any optimal adversarial strategy must have a core explanation. Further, by Lemma 4.2, such a core explanation is δ-core optimal. Using a (usually) much smaller mixed integer linear program, we can find such an explanation. We can then find the optimal adversarial strategy in polynomial time using BUILD STRAT.

Though we do not know what δ is, we know it must be in the range $[1,k]$. Further, using a relaxation of the OPT-KSEP-IPC constraints for solving geospatial abduction problems (as presented in Chapter 2; see also[9]), we can easily obtain a lower bound tighter than 1 on δ. Hence, if we solve k such (most likely small) mixed-integer-linear programs, we are guaranteed that at least one of them must be a core explanation for an optimal adversarial strategy. We note that these k MILPs can be solved in parallel (and the following k instances of BUILD-STRAT can also be run in parallel as well). An easy comparison of the results of the parallel processes would be accomplished at the end. As L^* is likely to be significantly smaller than L, this could yield a significant reduction in complexity. Furthermore, various relaxations of this technique can be used (*e.g.*, only using one value of δ).

Example 4.14. Continuing from Example 4.13, where the cartel members are attempting to find an OAS to best position drug laboratories, suppose they used the relaxation of OPT-KSEP-IPC (from Chapter 2 - see also [9]) to obtain a lower bound on the cardinality of an explanation and found it to be 2. With $k = 3$, they would solve two MILPs of the form of Example 4.13—one with $\delta = 2$ and one with $\delta = 3$. The solution to the first MILP would set X_{34} and X_{38} both to 1 while the second MILP would set X_{34}, X_{38}, and X_{57} all to 1. As the expected adversarial detriment for both solutions is 0, they are both optimal and running BUILD-STRAT is not necessary. Either $\{p_{34}, p_{38}\}$ or $\{p_{34}, p_{38}, p_{57}\}$ can be returned as an OAS.

4.4 Finding a Counter-Adversary Strategy

The preceding section explains how an intelligent adversary can try to keep its "partner" locations associated with a given set of observations hidden. To do this, the adversary uses an explanation function distribution—but unfortunately, the agent may not know what this distribution is. The agent is thus confronted with the problem of creating a strategy to discover the adversary's strategy. When attempting to find an "optimal" strategy for the agent, we first need to understand what benefit each possible strategy brings to the agent. More formally, we use a special case of expected reward (Definition 4.2.2 from Section 4.10) defined as the agent's *expected benefit* below.

Definition 4.19 (Expected Agent Benefit). Given a reward function **rf** and explanation function distribution exfd, the *expected agent benefit* is the function $\text{EXB}^{\text{rf}} : 2^{\mathscr{S}} \times \textbf{EFD} \to [0,1]$ defined as follows:

$$\text{EXB}^{\text{rf}}(\mathscr{B}, \text{exfd}) = \sum_{\text{ex\_fcn} \in \textbf{EF}} \textbf{rf}(\text{ex\_fcn}(\mathscr{O}, k), \mathscr{B}) \cdot \text{exfd}(\text{ex\_fcn})$$

Suppose an agent uses an explanation function distribution exfd to estimate how the adversary is assigning probabilities to specific explanation functions. $\text{EXB}^{\text{rf}}(\mathscr{B}, \text{exfd})$ is computed by looking at each explanation function ex_fcn, identifying the probability of ex_fcn according to the explanation function distribution, and then finding the reward for the counter-adversary strategy $\mathscr{B}$ used by the agent if ex_fcn were really the explanation function used. The product of the probability of ex_fcn and the reward to the agent of the counter-adversary strategy $\mathscr{B}$ yields an "expected reward" if ex_fcn is the actual explanation function—the sum of such products across all possible explanation functions yields the total expected reward. Thus, this definition is exactly identical to that of expected adversary detriment—except that we now consider the agent instead of the adversary.

Example 4.15. Following from Examples 4.1 and 4.6, suppose police detectives have information (*e.g.,* from a tipster) that the burglar is choosing safe locations according to exfd_{drug}. (Such information could also come from multiple runs of the GREEDY-KSEP-OPT2 algorithm of Chapter 2 (see also [9]). The police detectives wish to consider the set $\mathscr{B} = \{p_{41}, p_{52}\}$. First, they must calculate the reward associated with each explanation function (note that $k = 3, dist = 100$ and $\mathbf{rf} = \mathbf{crf}$).

$$\mathbf{crf}^{dist}(\text{ex\_fcn}_1(\mathscr{O}, 3), \{p_{41}, p_{52}\}) = 0.67$$
$$\mathbf{crf}^{dist}(\text{ex\_fcn}_2(\mathscr{O}, 3), \{p_{41}, p_{52}\}) = 0.5$$

(As an aside, we would like to point out the asymmetry in **crf**—compare these computations with the results of Example 4.7). Hence, $\text{EXB}^{\mathbf{crf}}(\{p_{41}, p_{52}\}, \text{exfd}_{drug}) = 0.634$.

We now define a counter-adversary strategy that the agent can use to nullify the agent's behavior with maximal effectiveness.

Definition 4.20 (Maximal Counter-Adversary Strategy (MCA)). Given a reward function **rf** and explanation function distribution exfd, a **maximal counter-adversary strategy**, $\mathscr{B}$, is a subset of $\mathscr{S}$ such that $\text{EXB}^{\mathbf{rf}}(\mathscr{B}, \text{exfd})$ is maximized.

Simply put, the maximal counter-adversary strategy is merely any strategy that yields the highest expected benefit to the agent. Note that in theory, there could be zero, one, or many potential maximal counter-adversary strategies.

Note that MCA does not include a cardinality constraint. This is because we do not require reward functions to be monotonic. In the monotonic case, we can trivially return all feasible points in $\mathscr{S}$ and be assured of a solution that maximizes the expected agent benefit. Therefore, for the monotonic case, we include an extra parameter $B \in \{1, \ldots, |\mathscr{S}|\}$ (for "budget") which will serve as a cardinality requirement for $\mathscr{B}$. This cardinality requirement for $\mathscr{B}$ is not necessarily the same as for $\mathscr{A}$ as the agent and adversary may have different sets of resources. Also, we do not require that $\mathscr{B}$ be an explanation. For a discussion of the special case where the solution to the MCA problem is required to be an explanation, see the appendix to [12].

4.4.1 The Complexity of Finding a Maximal Counter-Adversary Strategy

In this section, we develop complexity results for finding a maximal counter-adversary strategy. We start by formally defining the problem of finding a maximal counter-adversary strategy.

MCA Problem
INPUT: Space $\mathscr{S}$, feasibility predicate feas, real numbers α, β, set of observations $\mathscr{O}$, natural numbers k, B, reward function **rf**, and explanation function distribution exfd.
OUTPUT: Maximal counter-adversary strategy $\mathscr{B}$.

The result below shows that MCA is NP-hard via a reduction of the GCD problem.

Theorem 4.6. *MCA is NP-hard.*

Proof. Consider an instance of GCD consisting of set of points P, integer b, and integer K. We construct an instance of MCA as follows:
CONSTRUCTION:

- Set $\mathscr{S}$ to be a grid large enough that all points in P are also points in $\mathscr{S}$. We will use M, N to denote the length and width of $\mathscr{S}$.
- feas$(p) =$ TRUE if and only if $p \in P$
- $\alpha = 0$, and $\beta = \sqrt{M^2 + N^2}$, $\mathscr{O} = P$, $k = K$, and $B = K$
- Let $\mathbf{rf}(\mathscr{E}_1, \mathscr{E}_2)$ be **crf** where $dist = b$.
- Let functions ex_fcn$_1, \ldots,$ ex_fcn$_{|P|}$ be explanation functions, with each ex_fcn$_i$ corresponding to a unique $p_i \in P$. Let ex_fcn$_i(\mathscr{O}, num) = \{p_i\}$ for all $num > 0$. Note that each p_i is an explanation for the set P as it is of cardinality $\leq k$, is feasible, and is guaranteed to be with $[\alpha, \beta]$ from all other points in P as $[\alpha, \beta] = [0, \sqrt{M^2 + N^2}]$
- Let exfd(ex_fcn$_i$) $= \frac{1}{|P|}$ for all i.

CLAIM 1: $\mathbf{crf}^{dist}(\{p_i\}, \mathscr{B}) = 1$ if and only if there exists $p' \in \mathscr{B}$ such that a disc of radius b (note $b = dist$) centered on p' covers p_i. $\mathbf{crf}^{dist}(\{p_i\}, \mathscr{B}) = 0$ if and only if there does not exist $p' \in \mathscr{B}$ such that a disc of radius b centered on p' covers p_i. Follows directly from the definition of **crf**.

CLAIM 2: If the expected agent benefit is 1, then for all i, $\mathbf{crf}^{dist}(\{p_i\}, \mathscr{B}) = 1$. Suppose, by way of contradiction, that the expected agent benefit is 1 and there exists some p_i such that $\mathbf{crf}^{dist}(\{p_i\}, \mathscr{B}) \neq 1$. Then, for a singleton set, $\mathbf{crf}^{dist}(\{p_i\}, \mathscr{B}) = 0$. Hence, for the ex_fcn$_i$ associated with p_i, $\mathbf{crf}^{dist}(\text{ex\_fcn}_i(\mathscr{O}), \mathscr{B}) = 0$. So, by the definition of expected agent benefit, it is not possible for the expected agent benefit to be 1—a contradiction.

CLAIM 3: If MCA returns an optimal counter-adversary strategy with an expected expected agent benefit of 1, then GCD must also return "yes."
Suppose, by way of contradiction, MCA returns a strategy with an expected agent benefit of 1 and the corresponding of GCD returns "no." Then there does not exist a K-sized cover for the points in P. However, the set $\mathscr{B}$ is of cardinality K and by Claims 1–2 covers all points in P. Hence, a contradiction.

CLAIM 4: If GCD return "yes" then MCA must return an optima counter-adversary strategy with an expected agent benefit of 1.
Suppose, by way of contradiction, GCD returns "yes" and MCA returns a an optimal strategy with an expected agent benefit < 1. However, by the answer to GCD, there must exist $P' \subseteq P$ of cardinality k that is within distance b of all points in P. Hence, for all i, $\mathbf{crf}^{dist}(\{p_i\}, \mathscr{B}) = 1$ (as $b = dist$). So, the expected agent benefit must also be 1. Hence, a contradiction.

Proof of theorem: Follows directly from Claims 3–4.

The result below follows immediately from the proof of Theorem 4.6 and shows that MCA is NP-hard even if the reward function is monotonic.

Corollary 4.2. *MCA is NP-hard even if the reward function is monotonic.*

Later, in Section 4.4.4, we also show that MCA can encode the NP-hard MAX-K-COVER problem [6] as well (which provides an alternate proof for NP-hardness of MCA). We now present the decision problem associated with MCA and show that it is NP-complete under reasonable conditions.

MCA-DEC
INPUT: Space $\mathscr{S}$, feasibility predicate feas, real numbers α, β, set of observations $\mathscr{O}$, natural numbers k, B, reward function **rf**, explanation function distribution exfd, and number $R \in [0, 1]$.
OUTPUT: Counter-adversary strategy $\mathscr{B}$ such that $\mathsf{EXB}^{\mathbf{rf}}(\mathscr{B}, \mathsf{exfd}) \geq R$.

The following result says that as long as the reward function can be evaluated in polynomial time, the **MAX-DEC** decision problem is NP-complete. We note that all the example reward functions we have presented in this chapter are all polynomially computable and hence, the result below applies to them.

Theorem 4.7. *MCA-DEC is NP-complete, provided the reward function can be evaluated in PTIME.*

Proof. CLAIM 1: Membership in NP.
Given an explanation, $\mathscr{B}$, we can evaluate it reward and if it is an explanation in PTIME.

CLAIM 2: MCA-DEC is NP-hard.
Follows directly from Theorem 4.6

4.4.2 The Complexity of Counting MCA Strategies

Not only is **MCA-DEC** NP-hard, under the same assumptions as above, the result below establishes that counting version of the problem is #P-complete. Moreover, it has no fully polynomial random approximation scheme.

Theorem 4.8. *Counting the number of strategies that provide a "yes" answer to MCA-DEC is #P-complete and has no fully polynomial randomized approximation scheme (FPRAS for short) unless NP=RP.*

Proof. Theorem 4.6 shows a parsimonious reduction from GCD to **MCA**. Hence, we can simply apply Lemma 2.1 and the statement follows.

Theorem 4.8 tells us that MCA may not have a unique solution. Therefore, setting up a mixed strategy across all MCAs to determine the "best response" to the MCA of an agent by an adversary would be an intractable problem. This mirrors the result we presented in the preceding section (Theorem 4.3, page 111).

4.4.3 MCA in the General Case: Exact and Approximate Algorithms

In this section, we first describe an exact algorithm to find a maximal counter-adversary strategy for the agent. In the case of the IED detection example, for instance, a maximal counter-adversary strategy would correspond to the best places for US forces to search for IED weapons caches, given the presence of an adversary who is trying to conceal the locations of his caches. As the results above show, computing MCA is intractable computationally. Therefore, in this chapter, we also develop approximation algorithms that the agent could use to find a maximal counter-adversary strategy in the general case. Note that throughout this section (as well as in Section 4.4.4), we assume that the same pre-processing for **OAS** is used (cf. Section 4.3.2). We use the symbol L to refer to the set of all possible partners.

An Exact Algorithm For MCA. A naive, exact, and straightforward approach to the MCA problem would simply consider all subsets of L and pick the one which maximizes the expected agent benefit. Obviously, this approach has a complexity $O(\sum_{i=0}^{|\mathscr{S}|} \binom{|L|}{i})$ and is not practical. This is unsurprising as we showed this to be an NP-complete problem.

Approximation in the General Case. Despite the impractical time complexity associated with an exact approach, it is possible to approximate MCA with guarantees—even in the general case. This is due to the fact that when exfd is fixed, the expected agent benefit is submodular.[5]

---

[5] Recall that a function $f : 2^X \to \mathbb{R}$ is *submodular* if and only if for all subsets $X_1 \subseteq X_2 \subseteq X$ and for all $x \notin X_2$, it is the case that $f(X_1 \cup \{x\}) - f(X_1) \geq f(X_2 \cup \{x\}) - f(X_2)$.

Algorithm 12 (MCA-LS)

INPUT: Reward function **rf**, set $\mathcal{O}$ of observations, explanation function distribution exfd, possible partner set L, real number $\varepsilon > 0$
OUTPUT: Set $\mathcal{B} \subset \mathcal{S}$

1. Set $\mathcal{B}^* = L$, for each $p_i \in \mathcal{B}^*$ let $inc_i = \mathsf{EXB}^{\mathbf{rf}}(\{p\}, \mathsf{exfd}) - \mathsf{EXB}^{\mathbf{rf}}(\emptyset, \mathsf{exfd})$.
2. Sort the p_i's in $\mathcal{B}^*$ from greatest to least by inc_i (*i.e.*, p_1 is the element with the greatest inc_i).
3. $\mathcal{B} = \{p_1\}$, $\mathcal{B}^* = \mathcal{B}^* - \{p_1\}$, $cur\_val = inc_1 + \mathsf{EXB}^{\mathbf{rf}}(\emptyset, \mathsf{exfd})$, $flag1 = \mathsf{true}$, $i = 2$
4. While $flag1$

    a. $new\_val = cur\_val + inc_i$
    b. If $new\_val > (1 + \frac{\varepsilon}{|L|^2}) \cdot cur\_val$ then

        i. If $\mathsf{EXB}^{\mathbf{rf}}(\mathcal{B} \cup \{p_i\}, \mathsf{exfd}) > (1 + \frac{\varepsilon}{|L|^2}) \cdot \mathsf{EXB}^{\mathbf{rf}}(\mathcal{B}, \mathsf{exfd})$ then:

        $\mathcal{B} = \mathcal{B} \cup \{p_i\}$, $\mathcal{B}^* = \mathcal{B}^* - \{p_i\}$, $cur\_val = \mathsf{EXB}^{\mathbf{rf}}(\mathcal{B} \cup \{p_i\}, \mathsf{exfd})$
    c. If $new\_val \leq (1 + \frac{\varepsilon}{|L|^2}) \cdot cur\_val$ **or** if p_i is the last element then

        i. $j = 1$, $flag2 = \mathsf{true}$, number each $p_j \in \mathcal{B}$
        ii. While $flag2$
            A. If $\mathsf{EXB}^{\mathbf{rf}}(\mathcal{B} - \{p_j\}, \mathsf{exfd}) > (1 + \frac{\varepsilon}{|L|^2}) \cdot \mathsf{EXB}^{\mathbf{rf}}(\mathcal{B}, \mathsf{exfd})$ then:

            $\mathcal{B} = \mathcal{B} - \{p_j\}$, $cur\_val = \mathsf{EXB}^{\mathbf{rf}}(\mathcal{B} - \{p_j\}, \mathsf{exfd})$
            For each $p_i \in \mathcal{B}^*$ let $inc_i = \mathsf{EXB}^{\mathbf{rf}}(\mathcal{B} \cup \{p_i\}, \mathsf{exfd}) - \mathsf{EXB}^{\mathbf{rf}}(\mathcal{B}, \mathsf{exfd})$.
            Sort the p_i's in $\mathcal{B}^*$ from greatest to least by inc_i
            $i = 0$, $flag2 = \mathsf{false}$
            B. Else,
            If p_j was the last element of $\mathcal{B}$ then set $flag1, flag2 = \mathsf{false}$
            Otherwise, $j{+}{+}$
    d. $i{+}{+}$

5. If $\mathsf{EXB}^{\mathbf{rf}}(L - \mathcal{B}, \mathsf{exfd}) > \mathsf{EXB}^{\mathbf{rf}}(\mathcal{B}, \mathsf{exfd})$ then set $\mathcal{B} = L - \mathcal{B}$
6. Return $\mathcal{B}$

Theorem 4.9. *For a fixed* $\mathcal{O}, k, \mathsf{exfd}$, *the expected agent benefit,* $\mathsf{EXB}^{\mathbf{rf}}(\mathcal{B}, \mathsf{exfd})$ *has the following properties:*

1. $\mathsf{EXB}^{\mathbf{rf}}(\mathcal{B}, \mathsf{exfd}) \in [0, 1]$
2. *For* $\mathcal{B} \subseteq \mathcal{B}'$ *and some point* $p \in \mathcal{S}$ *where* $p \notin \mathcal{B}'$, *the following is true:*

$$\mathsf{EXB}^{\mathbf{rf}}(\mathcal{B} \cup \{p\}, \mathsf{exfd}) - \mathsf{EXB}^{\mathbf{rf}}(\mathcal{B}, \mathsf{exfd}) \geq \mathsf{EXB}^{\mathbf{rf}}(\mathcal{B}' \cup \{p\}, \mathsf{exfd}) - \mathsf{EXB}^{\mathbf{rf}}(\mathcal{B}', \mathsf{exfd})$$

(*i.e., expected agent benefit is sub-modular for MCA*).

It follows immediately that MCA reduces to the maximization of a submodular function. We now present the MCA-LS algorithm that leverages this submodularity. The basic intuition behind MCA-LS is the following.

1. Start with the set of all possible partners (the set L) and for each possible partners location p_i, find the *difference* of the expected benefit that occurs if we choose partner p_i to put in the agent's strategy as compared to not putting it in the agent's strategy. This value is the "incremental benefit" of adding p_i to the agent's strategy (when the agent's strategy is empty) and is denoted by inc_i. This is what happens in Line 1 of the MCA-LS algorithm.

2. In Line 2 of the MCA-LS algorithm, we sort L in descending order of the inc_i values.
3. In Line 3, we put the p_i with the highest inc_i value into the agent's strategy and remove it from consideration.
4. From Line 4 onwards, we execute a loop. In each iteration, we consider the next possible partner p_i from L—this is always the partner with the highest possible incremental benefit. In Line 4(a), we compute the new value of the agent's strategy if the incremental value of p_i can be directly added to the agent's strategy (the sum of the agent's strategy value plus inc_i is just an estimate). If this value is large enough (Line 4(b)), we add it to the agent's strategy; otherwise we do not add it.
5. The process is repeated several times until we are done.

We will explain what is meant by "large enough" via an example shortly.

The following two propositions leverage Theorem 4.9 and Theorem 3.4 of [5].

Proposition 4.15. *MCA-LS has time complexity of* $O(\frac{1}{\varepsilon} \cdot |L|^3 \cdot F(exfd) \cdot \lg(|L|))$ *where* $F(exfd)$ *is the time complexity to compute* $EXB^{rf}(\mathscr{B}, exfd)$ *for some set* $\mathscr{B} \subseteq L$.

Proof. We note that one iteration of the algorithm requires $O(|L| \cdot F(exfd) + |L| \cdot \lg(|L|))$ time. We shall assume that $O(|L| \cdot F(exfd)$ dominates $O(|L| \cdot \lg(|L|))$. By Theorem 3.4 of [5], the number of iterations of the algorithm is bounded by $O(\frac{1}{\varepsilon} \cdot |L|^2 \cdot \lg(|L|))$ where $F(exfd)$, hence the statement follows.

The result below now states that MCA-LS is a $\frac{1}{3}$-approximation algorithm for MCA and thus provides approximation guarantees.

Proposition 4.16. *MCA-LS is an* $(\frac{1}{3} - \frac{\varepsilon}{|L|})$-*approximation algorithm for MCA.*

Proof. By Theorem 4.9, we can be assured that when the "if" statement at line 4c is TRUE, then there are no further elements in $\mathscr{B}^*$ that will afford an incremental increase of $> (1 + \frac{\varepsilon}{|L|^2}) \cdot EXB^{rf}(\mathscr{B}, exfd)$, even if the last element is not yet reached. Hence, we can apply Theorem 3.4 of [5] and the statement follows.

We now return to our burglary example and use it to illustrate the running of our MCA-LS example.

Example 4.16. Let us consider our running example where law enforcement agents are attempting to find where a burglar resides in the area depicted in Figure 2.4. The agents may guess that there are at most k locations where the burglar might dwell (*e.g.*, home, office, significant other's house) and they know that these locations somehow support the burglaries that he carries out (set of observations $\mathscr{O}$). Furthermore, police assume that the burglar chose his safe locations using some explanation function distribution $exfd_{burglar}$ (see Example 4.6, page 107).

The law enforcement agents wish to find a maximal counter-adversarial strategy using the **prf** reward function (see page 4.2). They decide to use MCA-LS to find such a strategy with $\varepsilon = 0.1$. Initially (at line 3), the algorithm selects point p_{48} (renumbering as p_1, note that in this example we shall use p_i and inc_i numbering based on Example 2.5 rather than what the algorithm uses). Hence, $inc_{40} = 0.208$ and $cur\_val = 0.708$. As the elements are sorted, the next point to be considered in the loop at line 4 is p_{40} which has an incremental increase of 0, so it is not picked. It then proceeds to point p_{41}, which gives an incremental increase of 0.084 and is added to $\mathscr{B}$ so $cur\_val = 0.792$. Point p_{45} is considered next, which gives an incremental increase of 0.208 and is picked, so now $cur\_val = 1.0$. The algorithm then considers point p_{46}, which does not afford any incremental increase. After considering points $p_{33}, p_{35}, p_{37}, p_{42}, p_{43}, p_{44}, p_{47}, p_{49}, p_{50}, p_{52}, p_{56}$, and finding they all give a negative incremental increase (and thus, are not picked), the algorithm finds that the old incremental increase of the next element, p_1, would cause the "if" statement at line 4c to be true, thus proceeding to the inner loop inside that "if" statement (line 4(c)iiA). This loop considers if the removal of any picked elements p_{48}, p_{41}, p_{45} would cause the expected agent benefit to increase. However, in this example, if any of the elements are removed, the expected agent benefit decreases. Hence, the boolean $flag1$ is set to false and the algorithm exits the outer loop. The algorithm then returns the set $\mathscr{B} = \{p_{48}, p_{41}, p_{45}\}$ which is optimal.

4.4.4 Finding a Maximal Counter-Adversary Strategy, the Monotonic Case

In the previous section we showed a $\frac{1}{3}$ approximate solution to MCA can be found in polynomial time *even without any monotonicity restriction*. In this section, we show that under the additional assumptions of monotonicity of reward functions, we can obtain a better 63% approximation ratio with a faster algorithm. Here, we also have the additional cardinality requirement of B for the set $\mathscr{B}$ (as described in Section 4.4). We first show that expected agent benefit is monotonic when the reward function is.

Corollary 4.3. *For a fixed* $\mathcal{O}, k, \mathsf{exfd}$, *if the reward function is monotonic, then the expected agent benefit,* $\mathsf{EXB^{rf}}(\mathscr{B}, \mathsf{exfd})$ *is also monotonic.*

Proof. The zero-starting aspect of expected agent benefit follows directly from the definitions of zero-starting and expected agent benefit.

Consider the definition of $\mathsf{EXB^{rf}}$:

$$\mathsf{EXB^{rf}}(\mathscr{B} \cup \{p\}, \mathsf{exfd}) - \mathsf{EXB^{rf}}(\mathscr{B}, \mathsf{exfd}) \geq \mathsf{EXB^{rf}}(\mathscr{B}' \cup \{p\}, \mathsf{exfd}) - \mathsf{EXB^{rf}}(\mathscr{B}', \mathsf{exfd})$$

As $\mathbf{rf}$ is monotonic by the statement, and exfd is fixed, $\mathsf{EXB^{rf}}$ is a positive linear combination of monotonic functions, so the statement follows.

Algorithm 13 (MCA-GREEDY-MONO)

INPUT: Monotonic reward function **rf**, set $\mathscr{O}$ of observations, real number $B > 0$, explanation function distribution **exfd**, possible partner set L, real number $\varepsilon > 0$

OUTPUT: Set $\mathscr{B} \subset \mathscr{S}$

1. Initialize $\mathscr{B} = \emptyset$ and $\mathscr{B}^* = L$
2. For each $p_i \in \mathscr{B}^*$, set $inc_i = 0$
3. Set $last\_val = \mathsf{EXB}^{\mathbf{rf}}(\mathscr{B}, \mathsf{exfd})$
4. While $|\mathscr{B}| \leq B$

    a. $p_{best} = \mathsf{null}$, $cur\_inc = 0$
    b. For each $p_i \in \mathscr{B}^*$, do the following
       i. If $inc_i < cur\_inc$, break loop and goto line 4c.
      ii. Let $inc_i = \mathsf{EXB}^{\mathbf{rf}}(\mathscr{B} \cup \{p\}, \mathsf{exfd}) - last\_val$
     iii. If $inc_i \geq cur\_inc$ then $cur\_inc = inc_i$ and $p_{best} = p$
    c. $\mathscr{B} = \mathscr{B} \cup \{p_{best}\}$, $\mathscr{B}^* = \mathscr{B}^* - \{p_{best}\}$
    d. Sort $\mathscr{B}^*$ in descending order by inc_i.
    e. Set $last\_val = \mathsf{EXB}^{\mathbf{rf}}(\mathscr{B}, \mathsf{exfd})$

5. Return $\mathscr{B}$

Thus, when we have a monotonic reward function, the MCA problem reduces to the maximization of a monotonic, normalized[6] submodular function with respect to a uniform matroid[7]—this is a direct consequence of Theorem 4.9 and Corollary 4.3. Therefore, we can leverage the result of [7], to develop the MCA-GREEDY-MONO algorithm below. We improve performance by including "lazy evaluation" using the intuition that the incremental increase caused by some point p at iteration i of the algorithm is greater than or equal to the increase caused by that point at a later iteration. As with MCA-LS, we also sort elements by the incremental increase, which may allow the algorithm to exit the inner-loop earlier. In most non-trivial instances of MCA, this additional sorting operation will not affect the complexity of the algorithm (*i.e.*, under the assumption that the time to compute $\mathsf{EXB}^{\mathbf{rf}}$ is greater than $\lg(|L|)$, we make this same assumption in MCA-LS as well).

The basic outline of the MCA-GREEDY-MONO algorithm is as follows. As in the case of MCA-LS, we first assume the agent uses the empty strategy (he hasn't decided what to search as yet). We compute the expected benefit to the agent of using this strategy. We then iteratively add partners to the agent's strategy till the agent's strategy reaches the requisite size B. The key part of the MCA-GREEDY-MONO algorithm is in how we decide which points to add to the agent's strategy. In each iteration of the loop, we consider all remaining members of L and find the one, which, if added to the agent's strategy, gives the highest incremental benefit. Thus member of L then gets added to the agent's strategy, and the procedure is iteratively repeated until the agent's strategy reaches the desired size.

The result below specifies the running time of the MCA-GREEDY-MONO algorithm.

---

[6] As we include zero-starting in our definition of monotonic.

[7] In our case, the uniform matroid consists of all subsets of L of size B or fewer.

Proposition 4.17. *The complexity of MCA-GREEDY-MONO is* $O(B \cdot |L| \cdot F(\textbf{exfd}))$ *where* $F(\textbf{exfd})$ *is the time complexity to compute* $\textbf{EXB}^{\textbf{rf}}(\mathscr{B}, \textbf{exfd})$ *for some set* $\mathscr{B} \subseteq L$ *of size B.*

Proof. The outer loop at line 4 iterates B times, the inner loop at line 4b iterates $O(|L|)$ times, and at each inner loop, at line 4(b)ii, the function $\textbf{EXB}^{\textbf{rf}}(\mathscr{B}, \textbf{exfd})$ is computed with cost $F(\textbf{exfd})$. There is an additional $O(|L| \cdot \lg(|L|))$ sorting operation after the inner loop which, under most non-trivial cases, is dominated by the $O(|L| \cdot F(\textbf{exfd}))$ cost of the loop. The statement follows.

The result below shows us that MCA-GREEDY-MONO provides a 0.63 approximation ration for MCA when the reward function is monotonic.

Corollary 4.4. *MCA-GREEDY-MONO is an* $(\frac{e}{e-1})$-*approximation algorithm for MCA (when the reward function is monotonic).*

Proof. We need a definition of the notion of "incremental increase" in our proof:

Definition 4.21. For a given $p_i \in L$ at some iteration j of the outer loop of GREEDY-MONO (the loop starting at line 4), the *incremental increase*, $inc_i^{(j)}$, is defined as follows:

$$inc_i^{(j)} = \textbf{EXB}^{\textbf{rf}}(\mathscr{B}^{(j-1)} \cup \{p_i\}, \mathscr{A}) - \textbf{EXB}^{\textbf{rf}}(\mathscr{B}^{(j-1)}, \mathscr{A})$$

Where $\mathscr{B}^{(j-1)}$ is the set of points in L selected by the algorithm after iteration $j-1$.

We now continue with the proof of Corollary 4.4.

CLAIM 1: For any given iteration j of GREEDY-MONO and any $p_i \in L$, $inc_i^{(j)} \geq inc_i^{(j+1)}$

By Definition 4.21, the statement of the proposition is equivalent to the following:

$$\textbf{EXB}^{\textbf{rf}}(\mathscr{B}^{(j-1)} \cup \{p_i\}, \mathscr{A}) - \textbf{EXB}^{\textbf{rf}}(\mathscr{B}^{(j-1)}, \mathscr{A}) \geq \textbf{EXB}^{\textbf{rf}}(\mathscr{B}^{(j)} \cup \{p_i\}, \mathscr{A}) - \textbf{EXB}^{\textbf{rf}}(\mathscr{B}^{(j)}, \mathscr{A})$$

Obviously, as $\mathscr{B}^{(j-1)} \subseteq \mathscr{B}^{(j)}$, this has to be true by the submodularity of $\textbf{EXB}^{\textbf{rf}}$, as proved in Theorem 4.9.

By Claim 1, we can be assured that any point not considered by the inner loop will not have a greater incremental increase than some point already considered in that loop. Hence, our algorithm provides the same result as the greedy algorithm of [7]. We know that the results of [7] state that a greedy algorithm for a non-decreasing, submodularity function F such that $F(\emptyset) = 0$ is a $\frac{e}{e-1}$ approximation algorithm for the associated maximization problem. Theorem 4.9 and Corollary 4.3 show that these properties hold for finding a maximal counter-adversary strategy when the reward function is monotonic. Hence, by [7], the statement follows.

In addition to the fact that MCA-GREEDY-MONO is an $(\frac{e}{e-1})$-approximation algorithm for MCA, it also provides the best possible approximation ratio unless $P = NP$. In particular, the following result shows that there is not other polynomial algorithm that can provide an approximation ration which is strictly better than $(\frac{e}{e-1})$ unless $P = NP$. This is done by a reduction of MAX-K-COVER [6].

Theorem 4.10. *MCA-GREEDY-MONO provides the best approximation ratio for MCA (when the reward function is monotonic) unless $P = NP$.*

Proof. The MAX-K-COVER [6] is defined as follows.

INPUT: Set of elements, S and a family of subsets of S, $\mathcal{H} = \{H_1, \ldots, H_{max}\}$, and positive integer K.

OUTPUT: $\leq K$ subsets from $\mathcal{H}$ such that the union of the subsets covers a maximal number of elements in S.

In [6], the author proves that for any $\alpha' < \frac{e}{e-1}$, there is no α'-approximation algorithm for MAX-K-COVER unless $P = NP$. We show that an instance of MAX-K-COVER can be embedded into an instance of MCA where the reward function is monotonic and zero-starting in PTIME. By showing this, we can leverage the result of [6] and Corollary 4.4 to prove the statement. We shall define the reward function $srf(\mathcal{A}, \mathcal{B}) = 1$ if and only if $|\mathcal{A} \cap \mathcal{B}| \geq 1$ and $srf(\mathcal{A}, \mathcal{B}) = 0$ otherwise. Clearly, this reward function meets all the axioms, is zero-starting, and monotonic. We create a space $\mathcal{S}$ such that the number of points in $\mathcal{S}$ is greater than or equal to $|\mathcal{H}|$. For each subset in $\mathcal{H}$, we create an observation at some point in the space. We shall call this set $\mathcal{O}_{\mathcal{H}}$ and say that o_H is the element of $\mathcal{O}_{\mathcal{H}}$ that corresponds with set $H \in \mathcal{H}$. We set feas$(p) = $ true if and only if $p \in \mathcal{O}_{\mathcal{H}}$. We set $\alpha = 0$, β to be equal to the diagonal of the space, and $k = |\mathcal{O}_{\mathcal{H}}|$. Hence, any non-empty subset of $\mathcal{O}_{\mathcal{H}}$ is a valid explanation for $\mathcal{O}$. For each $x \in S$, we define explanation function ex_fcn$_x$ such that ex_fcn$_x(\mathcal{O}_{\mathcal{H}}, k) = \{o_H \in \mathcal{O}_{\mathcal{H}} | x \in H\}$. We define the explanation function distribution exfd to be a uniform distribution over all ex_fcn$_x$ explanation functions. We set the budget $B = K$. Clearly, this construction can be accomplished in PTIME. We note that any solution to this instance of MCA must be subset of $\mathcal{O}_{\mathcal{H}}$, for if it is not, we can get rid of the extra elements and have no change to the expected agent benefit. Hence, each $p \in \mathcal{B}$ will correspond to an element of $\mathcal{H}$, so we shall use the notation p_H to denote a point in the solution that corresponds with some $H \in \mathcal{H}$ (as each $o \in \mathcal{O}_{\mathcal{H}}$ corresponds with some $H \in \mathcal{H}$).

CLAIM 1: Given a solution $\mathcal{B}$ to MCA, the set $\{H \in \mathcal{H} | p_H \in \mathcal{B}\}$ is a solution to MAX-K-COVER.

Clearly, this solution meets the cardinality constraint, as there is exactly one element in $\mathcal{O}_{\mathcal{H}}$ for each element of $\mathcal{H}$ and $\mathcal{B}$ is a subset of $\mathcal{O}_{\mathcal{H}}$. Suppose, by way of contradiction, there is some other subset of $\mathcal{H}$ that covers more elements in S. Let $\mathcal{H}'$ be this solution to MAX-K-COVER and $\mathcal{B}'$ be the subset of $\mathcal{O}_{\mathcal{H}}$ that corresponds with it. We note that for some $x \in S$ in $\mathcal{B}'$, $srf(\text{ex\_fcn}_x(\mathcal{O}_{\mathcal{H}}, k), \mathcal{B}') = 1$ if and only if there is some $H \in \mathcal{H}'$ such that $x \in H$ and $srf(\text{ex\_fcn}_x(\mathcal{O}_{\mathcal{H}}, k), \mathcal{B}') = 0$ otherwise. Hence, the expected agent benefit is the fraction of elements in S covered by $\mathcal{H}'$. If $\mathcal{H}'$ is the optimal solution to MAX-K-COVER, then $\mathcal{B}'$ must provide a greater expected agent benefit than $\mathcal{B}$, which is clearly a contradiction.

CLAIM 2: Given a solution $\mathcal{H}'$ to MAX-K-COVER, the set $\{o_H \in \mathcal{O}_{\mathcal{H}} | H \in \mathcal{H}'\}$ is a solution to MCA.

Again, that the solution meets the cardinality requirement is trivial (mirrors that part of Claim 1). Suppose, by way of contradiction, there is some set $\mathcal{B}$ that provides a greater maximum benefit than $\{o_H \in \mathscr{O}_{\mathscr{H}} | H \in \mathscr{H}'\}$. Let $\mathscr{H}'' = \{H \in \mathscr{H} | p_H \in \mathscr{B}\}$. As with Claim 1, the expected agent benefit for $\mathscr{B}$ is equal to the fraction of elements in S covered by $\mathscr{H}''$, which is a contradiction as $\mathscr{H}'$ is an optimal solution to MAX-K-COVER.

The following example illustrates how MCA-GREEDY-MONO works.

Example 4.17. Consider the situation from Example 4.16, where the law enforcement agents are attempting to locate the burglar's places of residence. Suppose they want to locate these location, but use the **crf** reward function, which is monotonic (and hence also zero-starting). They use the cardinality requirement $B = 3$ in MCA-GREEDY-MONO. After the first iteration of the loop at Line 4, the algorithm selects point p_{48} as it affords an incremental increase of 0.417. On the second iteration, it selects point p_{46}, as it also affords an incremental increase of 0.417, so $last\_val = 0.834$. Once p_{46} is considered, the next point considered is p_{33}, which had a previous incremental increase (calculated in the first iteration) of 0.25, so the algorithm can correctly exit the loop to select the final element. On the last iteration of the outer loop, the algorithm selects point p_{35}, which gives an incremental increase of 0.166. Now the algorithm has a set of cardinality 3, so it exits the outer loop and returns the set $\mathscr{B} = \{p_{48}, p_{46}, p_{35}\}$, which provides an expected agent benefit of 1, which is optimal. Note that this would not be an optimal solution for the scenario in Example 4.16 which uses **prf** as p_{35} would incur a penalty (which it does not when using **crf** as in this example).

4.5 Implementation and Experiments

In this section, we describe prototype implementations and experiments for solving the OAS and MCA problems. For OAS, we create a MILP for the **crf** case and reduce the number of variables with the techniques we presented in Section 4.3. For MCA, we implement both the MCA-LS and MCA-GREEDY-MONO.

We carried out all experiments for MCA on an Intel Core2 Q6600 processor running at 2.4GHz with 8GB of memory available, using code written in Java 1.6; all runs were performed in Windows 7 Ultimate 64-bit using a 64-bit JVM, and made use of a single core. We also used functionality from the previously-implemented SCARE software from Chapter 2 to calculate, for example, the set of all possible partners L and to perform pre-processing (see the discussion in Section 4.3.2, page 111).

Our experiments are based on 21 months of real-world Improvised Explosive Device (IED) attacks in Baghdad[8] (see Chapter 2). The IED attacks in this 25×27 km region constitute our observations. The data also includes locations of caches associated with those attacks discovered by US forces. These constitute partner locations. We used data from the International Medical Corps to define feasibility predicates

---

[8] Attack and cache location data provided by the Institute for the Study of War.

based on ethnic makeup, location of US bases, and geographic features. We overlaid a grid of 100m × 100m cells—about the size of a standard US city block. We split the data into two parts; the first 7 months of data were used as a "training" set to learn the $[\alpha, \beta]$ parameters and the next 14 months of data were used for the observations. We created an explanation function distribution based on multiple runs of GREEDY-KSEP-OPT2 algorithm described in Chapter 2.

4.5.1 OAS Implementation

We now present experimental results for the version of OAS, with the **crf** reward function, based on the constraints in Definition 4.13 and variable-reduction techniques of Section 4.3.4. First, we discuss promising real-world results for the calculation of the reduced partner set L^*, described in Definition 4.15. Then, we show that an optimal adversarial strategy can be computed quite tractably using the methods discussed in Section 4.3.4. Finally, we compare our results to a set of real-world data, showing a significant decrease in the adversary's expected detriment across various parameter settings. Our implementation was written on top of the QSopt[9] MILP solver and used 900 lines of Java code.

Reduced Partner Set. As discussed in Section 4.3.2, producing an optimal adversarial strategy for any reward function relies heavily on efficiently solving a (provably worst-case intractable) integer linear program. The number of integer variables in these programs is based solely on the size of the partner set L; as such, the *ability to experimentally solve OAS* relies heavily on the size of this set.

Our real-world data created a partner set L with cardinality 22,692. We then applied the method from Definition 4.15 to reduce this original set L to a smaller subset of possible partners L^*, while retaining the optimality of the final solution. This simple procedure, while dependent on the explanation function distribution exfd as well as the cutoff distance for **crf**, always returned a reduced partner set L^* with cardinality between 64 and 81. This represents around a 99.6% decrease in the number of variables required in the subsequent integer linear programs!

Figure 4.4 provides more detailed accuracy and timing results for this reduction. Most importantly, regardless of parameters chosen, our real-world data is reduced by orders of magnitude across the board. Of note, we see a slight increase in the size of the reduced set L^* as the size of the explanation function distribution exfd increases. This can be traced back to the strict inequality in Definition 4.17. As we increase the number of nontrivial explanation functions in exfd, the number of nonzero constants $const_i$ increases. This results in a higher number of candidates for the intermediary set L^{**}. We see a similar result as we increase the penalizing cutoff distance. Again, this is a factor of the strict inequality in Definition 4.17 in conjunction with a higher fraction of nonzero $const_i$ constants.

---

[9] http://www2.isye.gatech.edu/~wcook/qsopt/index.html

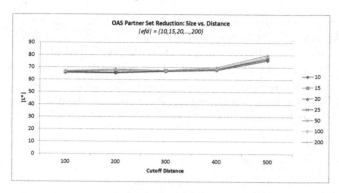

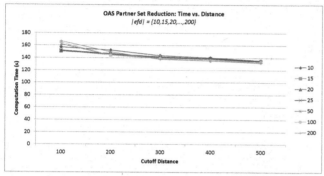

Fig. 4.4 The size of the reduced partner set L^* (top) and the time required to compute this reduction (bottom). Regardless of parameters chosen, we see a 99.6% decrease in possible partners—as well as integer variables in our linear program—in under 3 minutes.

Interestingly, Figure 4.4 shows a slight *decrease* in the runtime of the reduction as we increase the penalizing cutoff distance. Initially, this seems counterintuitive; with more nontrivial constants $const_i$, the construction of the intermediary set L^{**} requires more work. However, this extra work pays off during the computation of the final reduced set L^*. In our experiments, the reduction from L to L^{**} took less time than the final reduction from L^{**} to L^*. This is due to frequent short circuiting in the computation of the right-hand side of the conjunction during L^{**} creation. As we increase the penalizing cutoff distance, the size of L^{**} actually *decreases*, resulted in a decrease in the longer computation of L^*. As seen above, this decrease in L^{**} did not correspond to a decrease in the size of L^*.

Optimal Adversarial Strategy. Using the set L^*, we now present results to find an optimal adversarial strategy using δ-core optimal explanations. This is done by minimizing the MILP of Section 4.3.4, then feeding this solution into BUILD-STRAT. Since we do not know the value of δ in advance, we must perform this combined operation multiple times, choosing the best—lowest expected detriment—adversarial strategy as optimal.

A note on the lower bound for δ: as shown by [8], finding a *minimum-cardinality* explanation is NP-hard. Because of this, it is computationally difficult to find a tight lower bound for δ. However, this lower bound can be estimated empirically. For instance, for our set of real-world data from Baghdad, an explanation of cardinality below 14 has never been returned—even across tens of thousands of runs of GREEDY-KSEP-OPT2. Building on this strong empirical evidence, the minimum δ used in our experiments is 14.

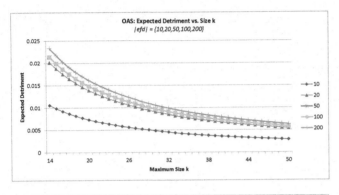

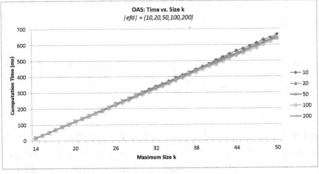

Fig. 4.5 Expected detriment of the optimal adversarial strategy (top) and the runtime of the integer linear program required to produce this strategy in milliseconds (bottom). Note the smooth decrease toward zero detriment as k increases, corresponding with a near-linear increase in total runtime.

Figure 4.5 shows both timing and expected detriment results as the size of the explanation function |exfd| and maximum strategy cardinality k are varied. Note that a lower expected detriment is better for the adversary, with zero representing no probability of partner discovery by the reasoning agent. As the adversary is allowed larger and larger strategies, its expected detriment smoothly decreases toward zero. Intuitively, as the number of nontrivially-weighted explanation functions in exfd increases, the expected detriment increases as well. This is a side effect of a larger |exfd| allowing the reasoning agent to cover a larger swath of partner locations.

Recall that, as the maximum k increases, we must solve linear programs for each $\delta \in \{k_{low}, k\}$. This is mirrored in the timing results in Figure 4.5, which assumes $k_{low} = 14$. As k increases, we see a near linear increase in the total runtime of the set of integer programs. Due to the reduced set L^*, we are able to solve dozens of integer programs in less than 800ms; were we to use the unreduced partner set L, this would be intractable. Note that the runtime graph includes that of BUILD-STRAT which always ran in under sixteen milliseconds.

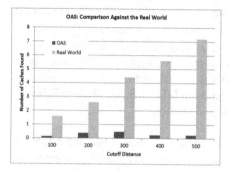 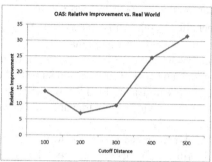

Fig. 4.6 Expected number of caches found when the adversary uses our strategy instead of the current state of the art (left - it is better for the adversary if fewer caches are found). Relative improvement of the OAS strategy versus the current state of the art (right). We assume the reasoning agent is using the Spatio-Cultural Abductive Reasoning Engine (SCARE) to provide information on cache locations.

OAS Performance w.r.t. Real-World Adversarial Strategy. Figure 4.6 compares the expected number of caches found under the current state of the art—IED cache locations based on 21 months of real-world data from Baghdad, Iraq—against the OAS strategy proposed in this paper. We hold the cardinality of the adversary's solution (*i.e.*, the number of possible caches) to 14 to match the real-world data. We assume the reasoning agent uses the Spatial Cultural Abductive Reasoning Engine (SCARE) introduced in [8] to provide partner locations to these attacks. SCARE is the state of the art method for finding IED caches.

When tested against real-world adversaries based on real-world Baghdad data, OAS significantly outperforms what adversaries have done so far in the real-world (fortunately this is balanced by later experiment results showing that MCA-LS and MCA-GREEDY-MONO significantly outperform SCARE). The expected number of caches found by SCARE against an opponent using OAS is significantly lower than against present day insurgents in Iraq. For instance, while SCARE (using a cutoff distance of 100 meters) detects 1.6 of the 14 possible caches against a real-world adversary, it is expected to detect only 0.11 of the caches against an adversary using OAS. This roughly order of magnitude improvement is seen across all five cutoff distances, from a minimum of approximately 7x at a cutoff distance of 200m to a maximum of over 31x at a distance of 500m. Thus, OAS significantly improves the adversary's performance.

4.5.2 MCA Implementation

First, we briefly discuss an implementation of the naive MCA algorithm discussed in section 4.4.3. Next, we provide promising results for the MCA-LS algorithm using the **prf** reward function. Finally, we give results for the MCA-GREEDY-MONO using the monotonic **crf** reward function, and qualitatively compare and contrast the results from both algorithms.

MCA-Naive. The naive, exact solution to MCA—considering all subsets of L with cardinality $k_{\mathscr{B}}$ or more and picking the one which maximizes the expected agent benefit—is inherently intractable. This approach has a complexity $O(\binom{|L|}{k_{\mathscr{B}}})$, and is made worse by the large magnitude of the set L. In our experimental setup, we typically saw $|L| > 20,000$; as such, for even the trivially small $k_{\mathscr{B}} = 3$, we must enumerate and rank over a trillion subsets. For any realistic value of $k_{\mathscr{B}}$, this approach is simply unusable. Luckily, we will see that both MCA-LS and MCA-GREEDY-MONO provide highly tractable and accurate alternatives.

MCA-LS. In sharp contrast to the naive algorithm described above, the MCA-LS algorithm provides (lower-)bounded approximate results in a tractable manner. Interestingly, even though MCA-LS is an approximation algorithm, in our experiments on real-world data from Baghdad using the **prf** reward function, the algorithm returned strategies with an expected benefit of 1.0 on every run. Put simply, on our practical test data, MCA-LS always *completely maximized* the expected benefit. This significantly outperforms the lower-bound approximation ratio of $1/3$. We would also like to point out that this is the first implementation (to the best of our knowledge) of the non-monotonic submodular maximization approximation algorithm of [5].

Since the expected benefit was maximal for every strategy $\mathscr{B}$ returned, we move to analyzing the particular structure of these strategies. Figure 4.7 shows a relationship between the size $|\mathscr{B}|$, the cutoff distance *dist*, and the cardinality of the expectation function distribution $|\mathsf{exfd}|$. Recall that **prf** penalizes any strategy that does not completely cover its input set of observations; as such, intuitively, we see that MCA-LS returns larger strategies as the penalizing cutoff distance decreases. If the algorithm can cover all possible partners across all expectation functions, it will not receive any penalty. Still, even when *dist* is 100m, the algorithm returns $\mathscr{B}$ only roughly twice the size as minimum-sized explanation found by GREEDY-KSEP-OPT2 (which, based on the analysis of Chapter 2 and [9], is very close to the minimum possible explanation). As the cutoff *dist* increases, the algorithm returns strategies with sizes converging, generally, to a baseline—the smallest-sized explanation found by the algorithm of [9], $|\mathscr{E}|$. This is an intuitive soft lower bound; given enough leeway from a large distance *dist*, a single point will cover all expected partners. This is not a strict lower bound in that, given two extremely close observations with similar expected partners, a single point may sufficiently cover both.

In Figure 4.8, we see results comparing overall computation time to both the distance *dist* and the cardinality of **exfd**. For more strict (*i.e.*, smaller) values of

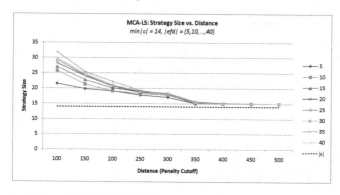

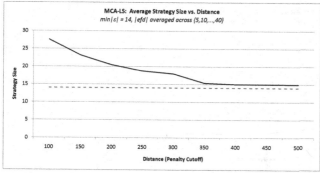

Fig. 4.7 The average size of the strategy recommended by MCA-LS decreases as the distance cut-off increases. For these experiments, the minimum cardinality for a given explanation $\mathscr{E}$ considered is exfd was 14, which gives us a natural lower bound on the expected size of a strategy. Note the convergence to this bound at cutoff distances at and above 300 meters.

dist, the algorithm—which, under **prf**, is penalized for all uncovered observations across exfd—must spend more time forming a strategy $\mathscr{B}$ that minimizes penalization. Similarly, as the distance constraint is loosened, the algorithm completes more quickly. Finally, an increase in $|\text{exfd}|$ results in higher computational cost; as explained in Proposition 4.15, this is due to an increase in $F(\text{exfd})$, the time complexity of computing $\text{EXB}^{\text{rf}}(\mathscr{B}, \text{exfd})$. Comparing these results to Figure 4.7, we see that the runtime of MCA-LS is correlated to the size of the returned strategy $\mathscr{B}$.

MCA-GREEDY-MONO. As discussed in Section 4.4.4, MCA-GREEDY-MONO provides tighter approximation bounds than MCA-LS at the cost of a more restrictive (monotonic) reward function. For these experiments, we used the monotonic reward function **crf**. Recall that a trivial solution to MCA given a monotonic reward function is $\mathscr{B} = L$; as such, MCA-GREEDY-MONO uses a budget B to limit the maximum size $|\mathscr{B}| \ll |L|$. We varied this parameter $B \in \{1, \dots, 28\}$.

Figure 4.9 shows the expected benefit $\text{EXB}^{\text{rf}}(\mathscr{B}, \text{exfd})$ increase as the maximum allowed $|\mathscr{B}|$ increases. In general, the expected benefit of $\mathscr{B}$ increases as the dis-

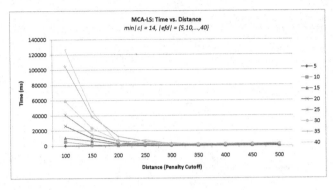

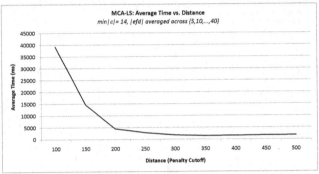

Fig. 4.8 The runtime of MCA-LS decreases as the penalizing cutoff distance is relaxed. Note the relation to Figure 4.7; intuitively, larger recommended strategies tend to take longer to compute.

tance constraint *dist* is relaxed. However, note the points with $B \in \{3, \ldots, 9\}$; we see that $dist \le 100$ performs better than $dist > 100$. We believe this is an artifact of our real-world data. Finally, as $|\text{exfd}|$ increases, the expected benefit of $\mathcal{B}$ converges more slowly to 1.0. This is intuitive, as a wider spread of possible partner positions will, in general, require a larger $|\mathcal{B}|$ to provide coverage.

Figure 4.10 shows that the runtime of MCA-GREEDY-MONO increases as predicted by Proposition 4.15. In detail, as we linearly increase budget B, we also linearly increase the runtime of our $F(\text{exfd}) = \text{EXB}^{\text{rf}}(\mathcal{B}, \text{exfd})$. In turn, the overall runtime $O(B \cdot |L| \cdot F(\text{exfd}))$ increases quadratically in B, for our specific reward function. Finally, note the increase in runtime as we increase $|\text{exfd}| = 10$ to $|\text{exfd}| = 100$. Theoretically, this increases $F(\text{exfd})$ linearly; in fact, we see almost exactly a ten-fold increase in runtime given a ten-fold increase in $|\text{exfd}|$.

MCA Algorithms and SCARE. We now compare the efficacy of the two MCA algorithms proposed in this paper to SCARE [8] which represents the current state of the art as far as IED cache detection is concerned. Again, our experiments are based on real-world data from Baghdad, Iraq. For these experiments, we average results across 100 runs of SCARE; as such, we hold $|\text{exfd}| = 100$ static for the MCA-based algorithms. Figure 4.11 plots the average number of predicted points within 500

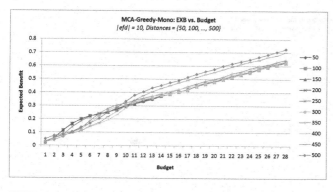

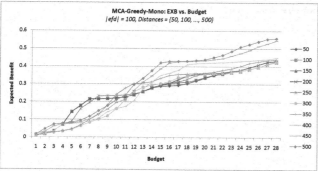

Fig. 4.9 Expected benefit of the strategy returned by MCA-GREEDY-MONO as the budget increases, with $|$exfd$| = 10$ (top) and $|$exfd$| = 100$ (bottom). Note the decrease in expected benefit due to the increase in $|$exfd$|$. Similarly, note the increase in expected benefit given a larger cutoff distance.

meters of an actual cache for both MCA-LS and MCA-GREEDY-MONO. SCARE, plotted as a horizontal line, predicts an average of 7.87 points within 500 meters of caches. MCA-LS finds over twice as many points at a low penalizing cutoff distances, and steadily converges to SCARE's baseline as the penalizing distance increases (as expected). As shown earlier in Figure 4.7, MCA-LS tends to find larger strategies given a smaller penalizing cutoff distance; in turn, these larger strategies yield more close points to actual caches. MCA-GREEDY-MONO shows similar behavior; as we increase the allowable budget (*i.e.*, maximum strategy size), more points are within 500 meters of a real-world cache location. Thus, MCA-LS and MCA-GREEDY-MONO both outperform SCARE, enabling more caches to be discovered.

We note that while the *number* of points in the strategy close to a real-world cache location is higher in the MCA-based algorithms than SCARE, the *fraction* of close points stays consistently close. SCARE returns a solution of size 14, with approximately half ($7.87/14 \approx 56\%$) of these points within 500 meters of cache. Compare this to, for instance, MCA-LS with a penalizing cutoff distance of 300

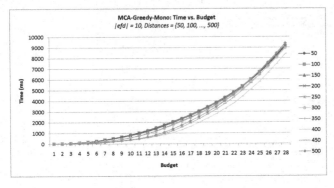

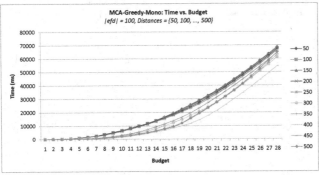

Fig. 4.10 Runtime of MCA-GREEDY-MONO as the budget increases, with $|\text{exfd}| = 10$ (top) and $|\text{exfd}| = 100$ (bottom). Note the increase in runtime due to the extra determinism of a larger exfd.

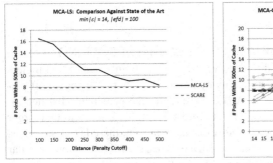

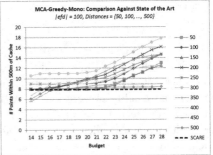

Fig. 4.11 Expected number of points within 500 meters of an actual cache returned by MCA-LS (left) and MCA-GREEDY-MONO (right) compared against an agent using SCARE (higher is better). Note that the SCARE software always returns an explanation of size 14, while both MCA algorithms benefit from the ability to adjust this explanation size.

meters; for these settings, the algorithm returns an average strategy size of 18, with 11 points (approximately 60%) within 500 meters of a cache location. This behavior is a product of the strategy size flexibility built into the MCA-based algorithms, and is beneficial to the reasoning agent. For example, assume the minimal solution to a problem is of size 2 and the reasoning agent has a budget of size 4. Now assume SCARE finds $1/2 = 50\%$ of the points near caches, while MCA-GREEDY-MONO finds $2/4 = 50\%$ of its points near caches. Both algorithms returned the same fraction of points near caches; however, the reasoning agent will spend its budget of 4 resources more effectively under MCA-GREEDY-MONO, instead of wasting 2 of its resources under the strategy provided by SCARE.

4.6 Conclusion

In this chapter, we recognized that adversaries are not going to sit by passively while the agent adapts to their behavior. Instead, the adversary is going to adapt its tactics in response to what the agent does as well. In our IED weapons cache detection application, for example, US forces observe what the adversary does, and use that information (using the techniques defined in Chapters 2 and 3) to determine which regions or locations to search for IED weapons caches. However, the work in those chapters assume that the adversary does not change his tactics, based on the searches that US forces carry out (that are very easily visible to them).

This chapter recognizes this reality and describes a mathematical framework, based on game theory, to determine how the adversary might adapt to his observations of the agent. We define this problem via notions of reward functions, leading to the definition of expected adversarial detriment (for the adversary). The adversary then tries to find a strategy that minimizes the expected adversarial detriment. For instance, in the case of the IED weapons cache location application, the adversary wants to find locations that minimize his expected adversarial detriment and, intuitively, minimizes the probability that his weapons cache locations will be found. We study the complexity of this problem and develop both exact and approximation algorithms to solve them.

The good news, for the agent, is that the adversary must move first. In the IED weapons cache detection application, the adversary must first decide where to put his weapons caches. The goal of the agent is to come up with a strategy (which corresponds to locations to search for weapons caches in the IED weapons cache detection application) which uncovers a maximal set of IED weapons caches. We formalize this problem in terms of expected benefit to the agent and find a strategy that maximizes the agent's expected benefit.

Our experiments to evaluate both the OAS algorithm to find an optimal adversary strategy and the MCA-Greedy-MONO algorithm to find the maximal counter-adversary strategy have been tested on real-world data involving IED attacks on US and Coalition forces in Iraq and have proven to be highly accurate.

References

1. Leyton-Brown, K., Shoham, Y. 2008. *Essentials of Game Theory: A Concise, Multidisciplinary Introduction.* Morgan and Claypool Publishers.
2. Johnson, D. 1982. The NP-Completeness Column: An Ongoing Guide. *Journal of Algorithms 3*, 2, 182–195.
3. Charnes, A., Cooper, W. 1962. Programming with linear fractional functionals. *Naval Research Logistics Quarterly 9*, 3, 163–297.
4. Karmarkar, N. 1984. A new polynomial-time algorithm for linear programming. *Combinatorica 4*, 4, 373–395.
5. Feige, U., Mirrokni, V. S., Vondrak, J. 2007. Maximizing non-monotone submodular functions. In *FOCS '07: Proceedings of the 48th Annual IEEE Symposium on Foundations of Computer Science.* IEEE Computer Society, Washington, DC, USA, 461–471.
6. Feige, U. 1998. A threshold of ln n for approximating set cover. *J. ACM 45*, 4, 634–652.
7. Nemhauser, G., Wolsey, L., Fisher, M. 1978. An analysis of the approximations for maximizing submodular set functions. *Mathematical Programming 14*, 265–294.
8. Shakarian, P., Subrahmanian, V.S., Sapino, M.L. SCARE: A Case Study with Baghdad, Proc. 2009 Intl. Conf. on Computational Cultural Dynamics (eds. D. Nau, A. Mannes), Dec. 2009, AAAI Press.
9. Shakarian, P., Subrahmanian, V.S., Sapino, M.L. 2012. GAPS: Geospatial Abduction Problems, ACM Transactions on Intelligent Systems and Technology (TIST), 3, 1, to appear.
10. Shakarian, P., Subrahmanian, V.S. Region-based Geospatial Abduction with Counter-IED Applications, accepted for publication in: Wiil, U.K. (ed.).Counterterrorism and Open Source Intelligence, Springer Verlag Lecture Notes on Social Networks, to appear, 2011.
11. Shakarian, P., Nagel, M., Schuetzle, B., Subrahmanian, V.S. 2011. Abductive Inference for Combat: Using SCARE-S2 to Find High-Value Targets in Afghanistan, in Proc. 2011 Intl. Conf. on Innovative Applications of Artificial Intelligence, Aug. 2011, AAAI Press.
12. Shakarian, P., Dickerson, J., Subrahmanian, V.S. 2012. Adversarial Geospatial Abduction Problems, ACM Transactions on Intelligent Systems and Technology (TIST), to appear.

Chapter 5
Two Real-World Geospatial Abduction Applications

Abstract In this chapter, we discuss two real-world applications of geospatial abduction problems (GAPs). While both applications deal with finding weapons caches that support improvised explosive devices used by insurgent terror groups, they operate in two different environments. SCARE (Spatio-Cultural Abductive Reasoning Engine) implements point-based geospatial abduction and was used to find IED weapons caches in Baghdad. In contrast, SCARE-S2 is an extension of SCARE to region-based geospatial abduction which has been used, with various modifications, to find high value targets (either large weapons caches or insurgent commanders) in certain provinces of Afghanistan. The accuracy of both systems has been tested on real-world data, and over 18 organizations have requested or used either SCARE or SCARE-S2.

5.1 Introduction

In this chapter, we describe the basic ideas behind two applications of geospatial abduction.

1. The *Spatio-Cultural Abductive Reasoning Engine* (SCARE [20]) implements point-based geospatial abduction as described in Chapter 2. SCARE tries to identify the locations of weapons caches in Baghdad. SCARE uses information about the cultural makeup of Baghdad, as well as information about the locations of natural features (*e.g.*, the Tigris river) and coalition bases, to define a feasibility predicate. SCARE was tested for accuracy using 21 months of real-world open source data about attacks in Baghdad (and about discovery of weapons caches in Baghdad)—7 months of data was used for training SCARE, while 14 months of data was used as a blind test data set to check accuracy.
2. The *SCARE-S2* system [23], on the other hand, has been applied to the problem of finding "high value targets" (or HVTs) in the Afghan provinces of Helmand and Kandahar. Again, using detailed information about the tribal geogra-

phy of these two provinces, we were able to define a feasibility predicate. Rather than Euclidean distances as used in SCARE, SCARE-S2 leveraged information about Afghan road networks to use "shortest path by road" distances. In addition, SCARE-S2 used region-based geospatial abduction as defined in Chapter 3 instead of point-based geospatial abduction. Finally, we used 6 months of real-world, open source data to train SCARE-S2 and to validate the accuracy of the system.

In the rest of this chapter, we describe the SCARE application (to IED cache detection) and SCARE-S2. We note that both systems can be applied to the other motivating examples presented earlier in this book if real-world data is available. This is apparent from the screenshots in the Introductory chapter which show SCARE being applied to examples such as the St. Paul, Minnesota church burglary scenarios, and the Tiger Detection application.

5.2 The Counter-IED Problem

The counterinsurgency environment provides a new set of challenges to the military commander, particularly at the tactical (Division, Brigade, Battalion, and lower) level. What von Clausewitz called the "fog of war" [1] is certainly present, but deceptive. Although the enemy in these contemporary conflicts often do not wear uniforms or operate out in the open, their actions in these complex environments are not entirely random. The enemy, or enemies, in a counterinsurgency typically have goals and strategies—not totally dissimilar to standard military units.

As with terrorist tactics, guerrilla tactics are neither mindless nor random. [2]

In the field of criminology, several theories exist that relate the geographic location of criminals with the locations of their crimes. *Pattern theory* [3] and *geographic profiling* [4] are extensively used. In the Army, intelligence professionals root their analysis in the process known as Intelligence Preparation of the Battlefield (IPB) [5, 25], which can also be extended to counter-insurgency operations [2]. However, traditionally, analysis of attacks in a counter-insurgency environment is to identify "hot spots" or places where attacks are likely to occur. In this chapter, we extend such analysis by examining techniques to locate sites used for enemy weapons caches based on attack data. We examine improvised explosive device (IED) attacks attributed to certain groups. We attempt to locate weapons cache sites based on attacks and on the locations of arrested enemy personnel using SCARE, the Spatial Cultural Abductive Reasoning Engine.

5.2.1 IEDs in an Insurgency

Though improvised explosive devices (IEDs) have been used for several decades dating from (at least) the time of the British military's presence in Ireland, they have emerged as a weapon of choice for the enemy in counter-insurgencies in Iraq and Afghanistan from 2001–2011. Until the development of SCARE, there were two main approaches to dealing with the IED problem. One approach is to focus on the attack: where the blast occurred, what type of explosive, etc. A common practice of commanders with this approach is to clear routes where IED attacks frequently occur and target IED networks through intelligence [6]. Another approach—to emphasize the IED "network"—is more intelligence focused and seeks to find bomb makers and emplacers [7].

Our approach with SCARE is a hybrid approach. We are using information about the attack to automatically create new intelligence about cache sites. These include information about feasibility predicates (*e.g.*, Shiite-backed attacks in Baghdad will not have weapons caches in either predominantly Sunni areas or on coalition bases, or in the middle of the Tigris river), as well as information about the $[\alpha, \beta]$ numbers that focus the search for IED weapons caches to donut-shaped regions centered at the location of an attack with the two concentric circles (of radii α, β respectively) defining the donut.

If uncovered, these cache sites can be exploited to gain further intelligence on the IED network through forensics and document exploitation. This will help lead to more effective counter-insurgency operations, by impeding the ability of the insurgent to transport and emplace the IEDs [8].

In order to use attack information to identify caches, we make some simplifying assumptions on the behavior of the IED attack cells. We know that IED attacks are typically conducted by small teams [9] whose members include the following:

- **IED manufacturers** who make the actual IED
- **IED emplacers** who place the IED in the designated attack area
- **IED triggermen** who are present during the IED attack. They may or may not arm or detonate the IED, but would at least conduct detailed surveillance of the attack
- **IED logisticians** ensure that IED manufacturers obtain materials or otherwise transport IEDs to and from cache sites
- **Higher level support** such as financial support, leadership, intelligence gathering, etc.

Social network analysis [10] is increasingly used to target IED networks. However, such analysis primarily focuses on higher-level support and IED manufacturers. On the other end of the spectrum, engagements with IED cell members at the location of attack will primarily target the emplacers and triggermen. SCARE will hit the logisticians and emplacers of the network as the caches are the key places where material is exchanged between the two. Furthermore, it has the potential to reduce the enemy's capability by denying them forward cache areas used for attacks. Without such cache sites, the IED cell members will be forced to travel longer distances with IED materials before an attack, thereby increasing the chances of compromise. The full list of assumptions for our model follows below.

1. IED cell members do not store materials at home. Typically, materials are stored in a common area (cache site).
2. The cache site is accessed prior to the attack to obtain the necessary materials.
3. The cell members have some restrictions on where the cache sites can be—*i.e.*, it cannot be in a body of water, on a coalition base, etc.
4. The distance from the cache site to the attack is greater than a certain distance α. If the cache site were too close, it would increase the chances of being found and destroyed following an engagement because US forces are likely to cordon off and search the immediate area in which an attack occurred.
5. The distance from the cache site to the attack must also be less than a certain distance β. Transporting munitions over too great a distance increases the chance of the cell members being compromised in transit (*i.e.*, material may only be moved during hours of darkness). For instance, a transporter may be stopped by a mobile checkpoint or by a random search team. Methods to reason about where such mobile checkpoints can and should be deployed have been developed in separate work by [11, 12] but we do not go into that topic in this book.

These assumptions have been commonly observed in combat enviroments, including Iraq and Vietnam [13].

Having this model of IED cell behavior is a starting point to creating an accurate representation of their behavior. We add a further constraint in that the attacks and cache sites are affiliated with the same insurgent group (or family of groups). The line of thinking is that different groups may use different models. Fortunately, we have open source data for the Iranian-backed "Special Groups" which conducted numerous IED operations in Iraq during 2007–2008.

5.2.2 Special Groups in Iraq

"Special Groups" operating in Iraq are defined as Shia extremist elements funded, trained, and armed by Iran [14, 15]. Although their influence seems to wax and wane over time and political situation [17], these groups leveraged significant insurgent military and political power during 2007–2008, and it is likely that US troop withdrawals will lead to a resurgence of activity by these groups.

Perhaps the most widely known among these groups is Jaysh al-Mahdi (JAM), headed by the young firebrand Shi'ite cleric Muqtada al-Sadr [18]. However, differing ideologies and agendas have caused fragmentation in this group. Despite the primacy of the Iraqi (Arab) Shi'ite identity that these groups publicly state, they receive a great deal of support from Iran [15] as cited by several sources. Many of the offshoot organizations of JAM also seem to retain the Iranian support as evidenced by their access to certain weapon systems.

The trademark weapon of these group is the explosively-formed penetrator (EFP). This weapon system is known to be imported from Iran [14]. This is a more advanced type of IED designed to penetrate armored vehicles. The signature of an EFP is sufficiently unique to the extent where it is easy to differentiate this type of attack from a typical IED in Iraq.

In Iraq, the Special Groups operate mainly in the southern provinces of Diyala, Salah-al-Din, and Baghdad [15]. In Baghdad, their safe havens are traditionally in districts of Sadr City and Kadamiyah. However, in 2006–2007, they attempted to exert their influence in other areas of Baghdad [19]. The Institute for the Study of War has developed a *Map of Special Groups Activity in Iraq* which we have used extensively. This map lists over 1000 insurgent activities attributed to Special Groups throughout Iraq. This data set contains events for the 21 months between February 2007 and November 2008, which is a period of high activity for these groups [16]. The events that are plotted are based on Multi-National Force – Iraq (MNF-I) press releases. According to the Institute for the Study of War, "efforts have been made to plot the data points with as much accuracy as possible."

The incidents in the map are only those attributed to the Special Groups. However, due to the nature of EFPs and militia affiliation, these events were relatively easy to identify with Special Groups with a high degree of accuracy. The activity types include the following categories:

1. Attacks with probable links to Special Groups
2. Discoveries of caches containing weapons associated with Special Groups
3. Detainments of suspected Special Groups criminals
4. Precision strikes against Special Groups personnel

In our tests, we utilize this map of Special Group activities as our data set. Next, we shall briefly describe the implemented SCARE system.

5.2.3 The SCARE System

In this section, we will describe how a user (such as a US soldier or a US commander) could interact with the SCARE system to identify weapons caches in some city. We do not recapitulate the theory of SCARE, as that has been well described in Chapter 2. Thus, we will merely go through the use of SCARE.

A user can bring up SCARE either on his local Google Maps-enabled computer or via a web portal. When he does so, he will see the screen shown in Figure 5.1 below.

This screen presents him with several tabs. He could use the demo data in the SCARE application which applies to Baghdad. Additionally, he could enter information in the Military Grid Reference System (MGRS) or in a Latitude/Longitude format. In all these cases, he needs to specify a rectangle referencing a rectangular region on the earth—the rectangle is specified through the coordinates of the lower left hand corner of the rectangle and the upper right hand corner of the rectangle.

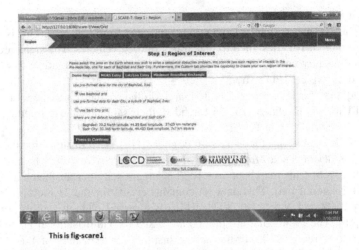

This is fig-scare1

Fig. 5.1 Main SCARE screen.

As mentioned earlier, these corners can be specified either via latitude/longitude coordinates or via MGRS coordinates as shown in Figure 5.2.

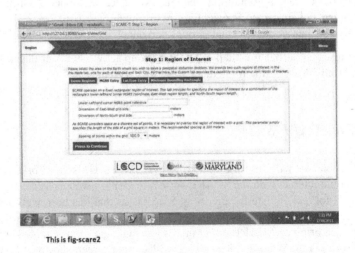

This is fig-scare2

Fig. 5.2 Main SCARE screen (using Military Grid Reference System, or MGRS).

Once this step is executed and the user has specified the region of the world in which he is interested (*e.g.*, a part of Baghdad or a part of Afghanistan or a tiger sanctuary in India), he can specify the distance constraints. These constraints specify the $[\alpha, \beta]$ values described in Chapter 2. The $[\alpha, \beta]$ values can either be learned from historical data or explicitly specified by the user. Thus, SCARE supports two possible use cases—one where the user does not want to guess $[\alpha, \beta]$ but wants the system to infer these values from historical data, and another where he thinks he can specify these values based on his knowledge of the application.

In the next two steps, shown in Figures 5.3 and 5.4, respectively, the user specifies the feasibility predicate. In step 3, the user specifies regions where the partner locations could be (these are called "support" zones) while in Step 4, the user can specify regions where the partner locations cannot possibly be (these are called "excluded zones"). In both cases, the user can specify these zones using Keyhole Markup Language (KML) files.

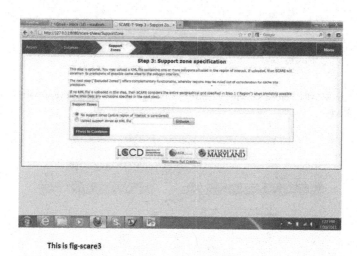

This is fig-scare3

Fig. 5.3 Main SCARE screen showing *support zones*, areas that could feasibly support partner locations.

In Step 5, SCARE allows the user to specify where the attacks took place. This can be done by uploading an Excel file with the information on where and when the attacks took place. This is shown in Figure 5.5.

Then, in step 6, the SCARE system asks the user to specify a number of iterations. Since the algorithm used by SCARE (see Chapter 2) is non-deterministic, the SCARE algorithm may make these non-deterministic choices a number of times, and this step of the SCARE system asks the analyst to specify the number of times the SCARE system should make these non-deterministic choices. This is shown in Figure 5.6.

Figure 5.7 shows the results screen generated by SCARE. This screen shows a histogram showing the caches predicted by the algorithms in Chapter 2, together with the number of attacks "supported" by each predicted cache (*i.e.*, the number of attacks within the stated $[\alpha, \beta]$ distance of each cache location). We can see, for example, in Figure 5.7, that caches at locations P15, P20 and P31 explain a very

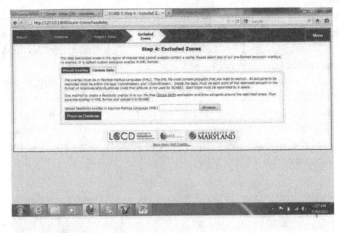

This is fig-scare4

Fig. 5.4 Main SCARE screen showing *exclusion zones*, areas that could never feasibly support a partner location.

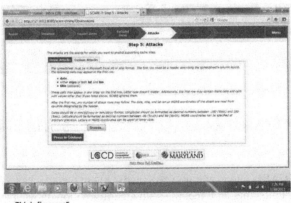

This is fig-scare5

Fig. 5.5 SCARE screen showing how to upload attack information, specifying where and when attacks occurred.

large number of attacks and so a commander might wish to deploy troops to look for IED weapons caches around those locations.

Figure 5.7 also contains links that allow us to see where the attacks occurred and where the predicted caches lie. These links correspond to the "Download Attacks as KML" and "Download Caches as KML" tabs shown in Figure 5.7. Selecting either

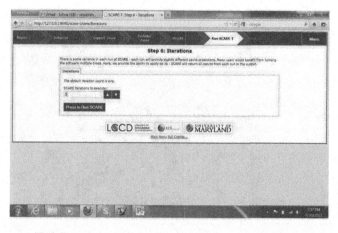

This is fig-scare6

Fig. 5.6 SCARE screen showing number of non-deterministic repetitions to be run.

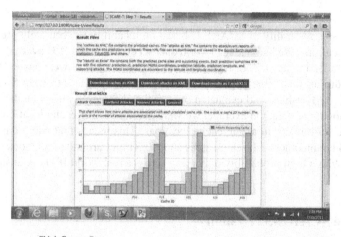

This is fig-scare7

Fig. 5.7 SCARE screen showing results. The histogram represents cache locations predicted by the algorithm over the set number of non-deterministic runs specified earlier.

of these tabs causes Google Earth to be invoked, showing the screen in Figure 5.8 below, mapping the predictions on their respective locations on the earth.

Thus, we see that SCARE allows users to seamlessly specify a number of parameters appropriate for their application. They can specify the part of the world in which they are interested, the feasibility predicate, can control the non-determinism inherent in the SCARE algorithm, and can visualize both the observations and the partners generated by SCARE using a Google Earth interface.

This is fig-scare8

Fig. 5.8 SCARE screen showing attacks and predicted caches using Google Earth.

Of course, the utility of SCARE is based on its ability to accurately predict the partner locations. We had one real-world data set consisting of IED attacks in Baghdad over a 21 month period, together with locations of IED weapons caches discovered during that same period. We used 7 months of data to train SCARE to learn the $[\alpha, \beta]$ values, and evaluated our predicted cache locations on the remaining 14 months of data. On average, we were able to predict cache locations within 0.45 miles (700m) from actual cache locations. This means that using our predictions, US commanders can search areas about 0.5 miles around predictions made by SCARE with a high probability of coming across a real world IED cache site.

5.3 The SCARE-S2 System

Unlike SCARE, the SCARE-S2 system focuses on helping find "High Value Targets" (HVTs) in the Helmand and Kandahar provinces of Afghanistan. Of course, SCARE-S2 can be applied to many other parts of Afghanistan, as well as many other parts of the world; however, we collected data pertinent to Helmand and Kandahar, and that is where we applied SCARE-S2. The work is described in detail in Chapter 3 and [23].

Insurgents operating in Afghanistan require substantial command-and-control (C2) and logistics support to conduct successful attacks. Military analysts refer to elements that provide C2 and logistics support for large number of insurgent cells as *high-value targets* (HVTs), as the elimination of these HVTs can have a significant impact on insurgent operations. As a result, NATO and Afghan forces often concentrate on finding these HVTs in an attempt to reduce the level of violence in the country. The insurgents have a limited number of these HVTs that are required to support the activities of lower-level insurgent cells. Additionally, terrain and cultural considerations place constraints on the relationships between an HVT and the lower-level insurgent cell it supports. Knowing the locations of the lower-level cells (based on attack data), as well as these constraints (obtained from socio-cultural and terrain data), we wish to abductively infer where the HVTs can be found. This is clearly an instance of a *geospatial abduction* problem originally introduced by the authors in [21] and later extended in [22], [24], and this book. We previously applied geospatial abduction to find small weapons hide-sites related to local attacks in Baghdad in [20]. However, the environment of Afghanistan provides several challenges that we did not address in Chapter 3. These include the following:

1. In Afghanistan, the influence of multiple tribes affect relationships between areas on the ground. How do we account for this influence?
2. In the two provinces we considered in Afghanistan, the terrain is extremely varied, unlike the more uniform urban terrain of Baghdad. How do we account for this variance in terrain?
3. Unlike our application to Baghdad (25×27 km area), where we could easily discretize the space, our data set for Afghanistan includes two provinces covering a total area 580×430 km, making discretization of the space impractical. How do we best represent the space?

We note that using only attack data and socio-cultural information alone will most likely be insufficient to pinpoint an HVT. However, the real-world requirements imposed on the insurgents by logistic and socio-cultural variables should allow a ground commander to significantly reduce the search space for such targets. Intelligence professionals identify *Named Areas of Interest* (NAIs), regions on the ground where they think HVTs can be located. Then, other intelligence assets such as unmanned aerial vehicles (UAVs) or tactical human intelligence (HUMINT) teams can be used in the NAIs to pinpoint targets [25]. In a large area, such as a province of Afghanistan, UAVs or HUMINT cannot be used effectively without first determining good NAIs. To address this problem for the specific case of Afghanistan, we adapted the region-based abduction framework of [22] (Chapter 3) to our scenario by creating an entirely new piece of software for abductive inference called SCARE-S2, the Spatio-Cultural Abductive Reasoning Engine – System 2.

SCARE-S2 abductively finds regions that can then later be used to cue other intelligence assets to find an HVT. Applying SCARE-S2 to our Afghanistan data set produced regions with a significantly higher density of HVTs (by a factor of 35), where half of the abduced ground regions (normally of an area less than 100km^2) would contain at least one HVT. Furthermore, each region produced by SCARE-S2 contained, on average, 4.8 villages—hence searching them is not resource-intensive for many surveillance platforms. Due to the high density of HVTs within the regions, we feel that they could be used for NAIs and aid in combat operations.

As region-based abduction has already been discussed in great detail in Chapter 3, we only focus on how the SCARE-S2 system works, rather than explaining, once again, how geospatial abduction works. As in the case of SCARE, SCARE-S2

can be be accessed either through the Internet or through a stand-alone installation
on a single computer.

5.3.1 SCARE-S2 Data Set

The data set used by SCARE-S2 consisted of HVTs and attack data from the Afghan
provinces of Helmand and Kandahar from January–June 2010, supported by tribal
and road network information. Below we provide details of the data set.

Provincial Data. All provincial data, including boundaries of provinces and districts, road networks, and village locations were provided by [26]. We considered
the Helmand and Kandahar provinces, which consist of 29 districts. The road network $(RN = (V,E))$ is an undirected graph of 30,304 vertices (1604 of which are
identified as villages) and 61,064 edges.

Attack Data. We used a series of 203 attack events in Afghanistan from [27]. 103
of these events were from January–April 2010 and were used to learn the $[\alpha,\beta]$
distance constraints, while the remaining 100 attacks (May–June 2010) were used
as set $\mathcal{O}$ of observations. We actually divided the set of observations into 12 subsets,
$\mathcal{O}_1 \subseteq \mathcal{O}_2 \subseteq \dots \subseteq \mathcal{O}_{12}$, with each subsequent set of observations containing 5 days
more worth of attacks than the previous (*i.e.*, $\mathcal{O}_1$ was May 2–6 and $\mathcal{O}_2$ was May
2–11). All attacks in the WITS database were identified by village—corresponding
to the AIMS information described earlier.

HVT Data. We collected a total of 78 HVTs based on official reports from [28].
These reports spanned January–September 2010. We used the reports of January–
April 2010 (27 HVTs) to learn the $[\alpha,\beta]$ distance constraints and the remainder for
a ground truth comparison. Notice that this time interval is greater than that used
for the set of observations, as an associated HVT with an attack may not necessarily
have been located in the same time window described earlier. As with the attack
data, each HVT was geo-located by the ISAF report with a village, which corre-
sponded to the AIMS information. We manually identified only certain weapons
caches and captured/killed enemy personnel as HVTs. Below we present our criteria in Figure 5.9. It is based on the combat experience of one of the authors.

Tribal Data. To create the *tribes* function, we used the tribal data from [29] that
associated districts in Afghanistan with a set of tribes. Altogether, there were 23
tribes reported by the NPS.

Distance Constraints. Using the simple algorithm FIND-BOUNDS (see [21] or
Chapter 2), which essentially returns an upper and lower distance bound on the
shortest distance to an HVT given a set of attacks, we determined the $[\alpha,\beta]$ bounds
to be $[0.0, 65.88]$ km based on the historical attack and HVT data from January–
April 2010.

- Cache HVTs:

  - The cache contains 3 or more mortar rounds
  - The cache contains mortar tubes
  - The cache contains 3 or more rockets
  - The cache contains 10 or more grenades
  - The cache contains 5 or more RPG launchers
  - The cache contains 20 or more RPG rounds
  - The cache contains 15 or more AK-47s (or other similar rifles)
  - The cache contains 3 or more land mines
  - The cache contains rooms full of communications equipment (or rooms full of any type of equipment)
  - The cache contains a DsHK or any other anti-aircraft weapon (including any number of Stinger missiles)
  - The cache contains 5 or more RPK machine guns (or similar capable systems such as M60, M249, etc.)
  - The cache contains 5 or more sniper rifles (such as a Dragunov)

- Personnel HVTs:

  - Reported listed individual as an insurgent "commander"
  - Reported listed individual as an insurgent "sub-commander"
  - Reported listed individual as an insurgent "planner"

Fig. 5.9 HVT criteria.

5.3.2 Adaptation of Region-Based Geospatial Abduction to Afghanistan

We used the region-based geospatial abduction method of Chapter 3 to capture geospatial abduction in Afghanistan.

There are two parts of the formalism of region-based abduction that are generally defined – the distance function (d) and the set of regions (R). In the experiments of [21] and Chapter 3, we used a Euclidean distance function and generated the regions from the REGION-GEN algorithm (see page 77), which discretizes the entire space—making it impractical for use here. Hence, we use d and R as a way to adapt region-based abduction to our Afghanistan scenario and address each of the three concerns outlined in the introduction. Our strategy is to build a special distance function, d_{afgh}, and use this function and the set of observations, $\mathcal{O}$, to generate R.

To address the first concern, that of multiple tribes, assume we have a set of tribes, $\mathcal{T} = \{\mathbf{t}_1, \ldots, \mathbf{t}_m\}$. Based on our data set, we can assume we have the following function $tribes : \mathcal{S} \to 2^{\mathcal{T}}$ which takes a point in the space and returns a set of tribes. Two points in the space, p_1, p_2, are *tribally-related* if and only if $tribes(p_1) \cap tribes(p_2) \neq \emptyset$. When we create our distance function, we will do so in a way to enforce this as an additional criterion that there must be at least one tribe that has a presence in the observation and partner location. The idea here is that an HVT must have a tribal-relationship with the lower-level cell conducting the attack, otherwise the two groups may not have a confluence of interest.

To address the second and third concerns, we appeal to the idea that the road networks of Afghanistan binds parts of this varied country together. Such sentiments are echoed in other work such as [30]. For any two villages on the road network (RN, an undirected graph where the vertices are villages) of Afghanistan, we define the function $sp_{RN} : \mathscr{S} \times \mathscr{S} \to \mathfrak{R}$ to return the shortest distance on the Afghan road network between the two points. Using the shortest path on a road network is also useful as our attack and HVT data were all geo-located by village. Hence, we put these concepts together to create our new distance function, d_{afgh}, defined below.

$$d_{afgh}(p,p') = \begin{cases} sp_{RN}(p,p') & \textit{iff tribes}(p) \cap \textit{tribes}(p') \neq \emptyset \\ \infty & \textit{otherwise} \end{cases}$$

We use this function to generate regions via the algorithm REGION-GEN-AFGH, presented below. A practical improvement we introduced was in determining the set V_o for each observation. We first determined the set V_o^{Euc}, which is V_o computed with a Euclidean distance function on the interval $[0, \beta]$—as the Euclidean distance function can be calculated much faster than shortest path. From this set, V_o is determined. It should be noted that the algorithm runs with a complexity $O(K \cdot |\mathcal{O}| \cdot T(RN))$ where K is a constant bound on the number of partners distance β away from a given observation and $T(RN)$ is the time complexity to find the shortest path between two points in RN. Another practical extension we added was to the output of GREEDY-REP-MC2. Any returned region over 1000 km^2 was not included in the output. Our intuition here is that a region so large is not useful to an analyst attempting to cue other intelligence assets.

REGION-GEN-AFGH

INPUT: Space $\mathscr{S}$, observations $\mathcal{O}$, reals α, β
OUTPUT: Set of regions R

1. Let the road-network, $RN = (V, E)$
2. For each $o \in \mathcal{O}$, find the set $V_o = \{v \in V | d_{afgh}(o, v) \in [\alpha, \beta]\}$
3. Let $L = \bigcup_{o \in \mathcal{O}} V_o$. For each $p \in L$ let $\mathcal{O}_p$ be the set of observations that can be associated with it.
4. Partition L into subsets, denoted $L_{\mathcal{O}'}$, where $\mathcal{O}' \subseteq \mathcal{O}$ and $p \in L_{\mathcal{O}'}$ iff $\mathcal{O}_p = \mathcal{O}'$.
5. For each $L_{\mathcal{O}'}$, create region r that is the minimum-enclosing rectangle of all elements in $L_{\mathcal{O}'}$. Add r to R.
6. Return set R.

5.3.3 SCARE-S2 Experiments

The experimental results for SCARE-S2 only focus on villages in Helmand and Kandahar province being feasible locations. Moreover, distances used to specify the $[\alpha, \beta]$ constraints were road distances in SCARE-S2 instead of Euclidean distance in SCARE. Finally, the feasibility predicate used by SCARE-S2 used the tribally-related function to define feasibility (*e.g.*, if a tribe was believed to have carried out an attack, the only feasible locations for a weapons cache supporting that attack were villages where the same tribe had a presence).

Setup. Our implementation of and corresponding experiments with SCARE-S2 ran on a Lenovo T400 ThinkPad laptop equipped with an Intel Core 2 Duo T9400 processor operating at 2.53 GHz and 4.0 GB of RAM. The computer was running Windows Vista 64-bit Business edition with Service Pack 1 installed. This modest hardware setup was selected as deploying units to Afghanistan are typically equipped with Windows-based laptop systems. Isolated command posts, with limited connectivity to a network due to terrain restrictions may only have access to this limited computational power.

We implemented SCARE in approximately 4000 lines of Java code. Java Runtime Environment (JRE) Version 6 Update 14 was used. The software was developed with Eclipse version 3.4.0. We used the JGraphT library version 0.81 to implement the Fibonacci heap and the graph structure. Additionally, BBN OpenMap was used for some of the geospatial methods. We also added the capability to output KML files so that the results could be viewed in Google Earth; we used Google Earth 4.3.7284.3916 (beta) operating in DirectX mode. Experimental results were also collected in CSV-formatted spreadsheets.

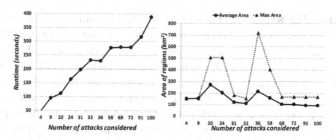

Fig. 5.10 Number of attacks versus runtime (average over 10 trials) and average region area.

Runtime Experiments. We examined runtime of the algorithm by running the algorithm on each of the 12 subsets of observations described earlier. We observed two things: that the relationship between runtime and number of attacks was linear and that the runtime of REGION-GEN-AFGH dominated the runtime of GREEDY-REP-MC2 (which was negligible). This is primarily the result of the calculation of the shortest path. As stated earlier, this relationship is linear, so our result depicted in Figure 5.10 is unsurprising.

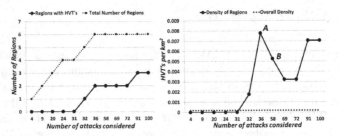

Fig. 5.11 Number of attacks versus number of regions and HVT density.

Area of Regions. As with Chapter 3 (also see [22]), we examine the average area of the regions. In general, smaller regions are preferred and as the set of observations $\mathcal{O}$ grows, the regions should become smaller. In each of the 12 trials, there was never more than one region over 200 sq km, and as set $\mathcal{O}$ increased, the average area approach 100 sq km. This is exactly what is desired. We plot the average and maximum areas in Figure 5.10. Note that a few spikes in average area are directly related to spikes in maximum area from a few outliers produced on some runs. Note that only a third of our runs produced a region over 200 sq km. Although even 100 sq km may seem like a large area, we must consider the density of villages, which is where we are attempting to locate caches. The overall density of villages for the entire area considered was 0.0064.[1] By the nature of how the regions are generated, they inherently have more villages. We observed that when we considered the entire set of attacks, no region contained more than 8 villages, with an average village density of 4.8 villages per region. As such, we feel that the regions produced by SCARE-S2 will be helpful in directing intelligence, surveillance, and reconnaissance (ISR) assets.

HVTs Enclosed by Regions. In Figure 5.11, we plot the number of regions returned by each run, as well as the number of regions that enclose at least one HVT from the ground truth set. Although the number of regions increases with the number of attacks (from 1 to 6), the number of regions enclosing an HVT also increases (from 0 to 3). While we should expect that solutions with more regions enclose more HVTs, we must also recall that the regions become smaller with each run. Furthermore, we also examined HVT density (number of HVTs divided by total area of all regions), which also increased with each run. Note we had two outliers, identified in Figure 5.11 as points A and B. In these two runs, the software returned larger regions of size 719.68 sq km and 403.34 sq km that enclosed a large urban area where many HVTs were found. Eliminating these regions from the solution would eliminate these artificial spikes in density. When we considered the entire two months of observed attacks, the HVT density in the regions was over 35 times greater than the overall HVT density in the provinces. We remind the reader that

---

[1] In the newest version, SCARE-S2 also runs the geospatial abduction algorithm of Chapter 2 (also see[21]) which abduces points (villages, in this case). Hence, the output now not only included regions, but villages of interest as well, which allows us to further reduce the search-space for HVTs.

the the regions are meant to be used as Named Areas of Interest (NAIs) for use by intelligence personnel. These NAIs would then be used to cue other intelligence assets (for example, a UAV or a HUMINT team) to conduct a more fine-grained search (avoiding a search in a larger area). Therefore, despite only half the returned SCARE regions containing NAIs, the small size of the regions, along with the high density of HVTs, make them invaluable for the intelligence process.

Discussion. We shall now consider our final run of the algorithm, where we considered the entire set of 100 attacks from May–June 2010. This run produced the most regions enclosing HVTs, the greatest HVT density (discounting spikes A and B), and the smallest average region area.

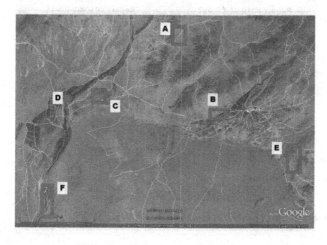

Fig. 5.12 Regions returned after considering attacks from May–June 2010.

This trial of the software produced 6 regions, labeled A–F, shown in Figure 5.12. Half of them enclosed an HVT. There were other ISAF reports that did not include village information. We did not consider these additional reports in any part of our experiments. However, all three regions returned by this experiment that did not enclose an HVT were located in districts where an HVT was reported (with no village information). For region D, there were 11 such reports, for region F, there were 4 such reports, and for district E there were was one such report. Let us now consider the HVTs found within regions A–C, depicted in Figure 5.13. Region A (with an area of 102.5 sq km) encloses the village of Bahram in the Ghorak district of Kandahar. According to ISAF PAO report 2010-05-CA-052, on May 5, 2010, a combined ISAF-Afghan force captured a Taliban commander in this village, who was responsible for several improvised explosive device (IED) attacks as well as movement of foreign fighters in the country. He also had a cache that included automatic rifles and heroin. Region B (with an area of 72.0 sq km) encloses the village of Makuan, in the Zhari district of Kandahar. According to ISAF PAO report 2010-07-CA-11, on July 18, 2010, a combined Afghan-ISAF force conducted a raid on a compound where a Taliban weapons facilitator was believed to reside. The unit received fire from insurgents, and returned fire killing several of them. As they approached the compound, they found several IEDs placed to guard the facility. The compound was found to be a IED factory and a bunker system that contained munitions. Region C (with an area of 71.0 sq km) encloses the village Kharotan in the Nahri Sarraj district of Hilmand. ISAF PAO report 2010-08-CA-161 describes how ISAF forces detained the Taliban deputy-commander of the Lashkar Gah district there on August 14, 2010.

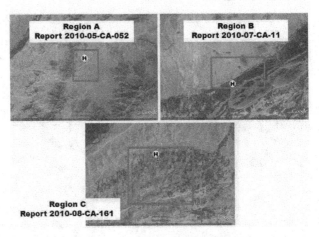

Fig. 5.13 Close-up view of regions A–C with actual HVTs plotted.

References

1. von Clausewitz, C. 1832. *On War*. Wilder Publications.
2. US Army. 2006. *Counterinsurgency (US Army Field Manual)*, FM 3-24 edition.
3. Brantingham, P., and Brantingham, P. 2008. Crime Pattern Theory. In Wortley, R., and Mazerolle, L., eds., *Enviromental Criminology and Crime Analysis*. 78–93.
4. Rossmo, D. K., and Rombouts, S. 2008. Geographic Profiling. In Wortley, R., and Mazerolle, L., eds., *Enviromental Criminology and Crime Analysis*. 136–149.
5. US Army. 2004. *Intelligence (US Army Field Manual)*, FM 2-0 edition.
6. Brown, R. A. 2007. Commander's Assessment: South Baghdad. *Military Review* 27–34.
7. Moulton, J. 2009. Rethinking IED Strategies from Iraq to Afghanistan. *Military Review*.
8. Mansoor, P. R., and Ulrich, M. S. 2007. A New COIN Center-of-Gravity Analysis. *Military Review*.
9. McFate, M. 2005. Iraq: The Social Context of IEDs. *Military Review*.
10. Reed, B. 2007. A Social Network Approach to Understanding an Insurgency. *Parameters*.
11. Dickerson, J., Simari, G., Subrahmanian, V., Kraus, S. 2010. A Graph-Theoretic Approach to Protect Static and Moving Targets from Adversaries. In *Proc. 9th Int. Conf. on Autonomous Agents and Multiagent Systems (AAMAS-2010)*. 299–306.

12. Paruchuri, P., Tambe, M., Ordóñez, F., Kraus, S. 2006. Security in multiagent systems by policy randomization. In *Proc. 5th Int. Conf. on Autonomous Agents and Multiagent Systems (AAMAS-2006)*.

13. Shakarian, P., Otstott, C. What is Old is New: Countering IEDs by Disrupting the Weapon Supply, Military Review, pp. 46–52, 2011.

14. Kagan, F.; Kagan, K.; and Pletka, D. 2008. Iranian Infulence in the Levant, Iraq, and Afghanistan. Technical report, American Enterprise Institute.

15. Cochrane, M. 2008c. The Growing Threat of Special Groups in Baghdad. Backgrounder 25, The Institute for the Study of War.

16. Cochrane, M. 2008b. Special Groups Regenerate. Iraq Report 11, The Institute for the Study of War.

17. Cochrane, M. 2009. The Fragmentation of the Sadrist Movement. Iraq Report 12, The Institute for the Study of War.

18. Nasr, V. 2007. *The Shia Revival*. W.W. Norton and Co.

19. IMC. 2007. IRAQIS ON THE MOVE: Sectarian Displacement in Baghdad, An Assessment by International Medical Corps.

20. Shakarian, P., Subrahmanian, V.S., Sapino, M.L. SCARE: A Case Study with Baghdad, Proc. 2009 Intl. Conf. on Computational Cultural Dynamics (eds. D. Nau, A. Mannes), Dec. 2009, AAAI Press.

21. Shakarian, P., Subrahmanian, V.S., Sapino, M.L. 2012. GAPS: Geospatial Abduction Problems, ACM Transactions on Intelligent Systems and Technology (TIST), 3, 1, to appear.

22. Shakarian, P., Subrahmanian, V.S. Region-based Geospatial Abduction with Counter-IED Applications, accepted for publication in: Wiil, U.K. (ed.).Counterterrorism and Open Source Intelligence, Springer Verlag Lecture Notes on Social Networks, to appear, 2011.

23. Shakarian, P., Nagel, M., Schuetzle, B., Subrahmanian, V.S. 2011. Abductive Inference for Combat: Using SCARE-S2 to Find High-Value Targets in Afghanistan, in Proc. 2011 Intl. Conf. on Innovative Applications of Artificial Intelligence, Aug. 2011, AAAI Press.

24. Shakarian, P., Dickerson, J., Subrahmanian, V.S. 2012. Adversarial Geospatial Abduction Problems, ACM Transactions on Intelligent Systems and Technology (TIST), to appear.

25. US Army. 1994. *Intelligence Preparation of the Battlefiled (US Army Field Manual)*, FM 34-130 edition.

26. Afghanistan Information Management Services (AIMS). GIS / Mapping Services, http://http://www.aims.org.af/.

27. National Counter-Terrorism Center (NCTC). Worldwide Incident Tracking System (WITS), https://wits.nctc.gov/.

28. International Security Assistance Force (ISAF) Afghanistan. Press Releases, http://www.isaf.nato.int/article/isaf-releases/index.php.

29. Naval Postgraduate School (NPS). Program for Culture and Conflict Studies, http://www.nps.edu/programs/ccs/.

30. Conover, T. 2010. *The Routes of Man: How Roads Are Changing the World and the Way We Live Today*. Knopf.

Index